THERE IS A CD-ROM DISK THAT
ACCOMPANIES THIS BOOK.

PLEASE ASK AT THE COUNTER
FOR FURTHER DETAILS

MRI Atlas
of the Human Cerebellum

MRI Atlas
of the Human Cerebellum

Jeremy D. Schmahmann

Department of Neurology, Massachusetts General Hospital
and Harvard Medical School
Boston, Massachusetts

Julien Doyon

Department of Psychology and Rehabilitation Research Group
François-Charon Center, Laval University
Quebec City, Quebec, Canada

Arthur W. Toga

Laboratory of Neuroimaging, Department of Neurology
University of California School of Medicine
Los Angeles, California

Michael Petrides

Cognitive Neuroscience Unit, Montreal Neurological Institute
and McGill University, Montreal, Quebec, Canada

Alan C. Evans

McConnell Brain Imaging Center, Montreal Neurological Institute
and McGill University, Montreal, Quebec, Canada

ACADEMIC PRESS
A Harcourt Science and Technology Company

San Diego   London   Boston   New York   Sydney   Tokyo   Toronto

Academic Press
A Harcourt Science and Technology Company
525 B Street, Suite 1900, San Diego, California 92101-4495, USA
http://www.academicpress.com

Academic Press
Harcourt Place, 32 Jamestown Road, London NW1 7BY, UK
http://www.hbuk.co.uk/ap/

Library of Congress Catalog Card Number: 00-100269

International Standard Book Number: 0-12-625665-9

PRINTED IN THE UNITED STATES OF AMERICA
00 01 02 03 04 05 PTP 9 8 7 6 5 4 3 2 1

To our
families and mentors.

Contents

FOREWORD   ix

PREFACE   xi

ACKNOWLEDGMENTS   xiii

Introductory Text   1

Images   21

Sagittal Series   21

Coronal Series   91

Horizontal Series   133

Foreword

THE ADVENT OF neuroimaging technology (CT, fMRI, PET, SPECT) has opened a new era of functional anatomy—one that might be called the era of *Bioanatomy*. It allows for the first time the observation of the activity of the normal or diseased human brain as it performs assigned tasks. Much that is new should come from the assiduous application of these methods.

The most widely recognized functions of the cerebellum relate to posture and volitional movement, which were well described by Babinski and Holmes and conceived by Sherrington as the activity mainly of the "head ganglion of the proprioceptive system." But the wide connectivity of the cerebellum with all parts of the cerebrum, including the auditory and visual cortices, belies this restricted view and suggests other possibilities.

In past times clinicopathological studies were the main sources of our knowledge of the cerebellum. They were the subjects of important articles by Dow and Moruzzi, A. Brodal and Jansen, and Greenfield. Important as the articles were, they did not describe the lesions in terms of the current anatomy of the cerebellum. The atlas of the normal human cerebellum compiled by Angevine, Mancall, and Yakovlev (1961) was intended to provide better maps to which the diseased organ could be compared. Victor, Mancall, and I used it in our study of the effects of alcohol on the cerebellum, as I pointed out in the preface to that atlas.

However, for reasons recounted on the following pages, the atlas of Angevine, Mancall, and Yakovlev, based solely on stained sections of postmortem specimens, proved to be unsuitable for contemporary bioanatomical studies. This was the main motivation for the preparation of this new atlas. Although most of it surveys the cerebellum of an adult Caucasian male, it is expected that the method will be extended to other individuals of different ages, sexes, and races.

The reader will find this atlas helpful in another way: It clarifies anatomical nomenclature. The literature on the cerebellum has been burdened with a number of archaic and occult terms. Words such as *declive, pyramis, simplex, nodulus, flocculus* and even *vermis* are obscure and are often used in different ways. The table of a new terminology, proposed by the authors, if adopted, will simplify and unify terminology in future publications.

Raymond D. Adams, M.D.
Bullard Professor of Neuropathology, Emeritus
Massachusetts General Hospital and Harvard Medical School
Boston, Massachusetts

Preface

SINCE THE FIRST detailed written description of the cerebellum by Vincenzo Malacarne in 1776 and the brain atlas of Felix Vicq-d'Azyr in 1786 that specifically depicted the structure and subdivisions of the cerebellum, there have been numerous attempts to display and describe the gross morphologic organization of the human cerebellum. This atlas, like its predecessors, reflects the evolving state of knowledge of the organization and function of the cerebellum and the contemporary technology available to investigate it. Anatomic and functional magnetic resonance imaging (fMRI) technology has evolved considerably in recent years, and the ability to differentiate cerebellar lobules and folia on all planes of section has become quite precise. There is a problem, however, in correlating this technology with the knowledge of cerebellar anatomy because there is no atlas that can be reliably used for the accurate identification of the different cerebellar regions. Terms such as medial/lateral, anterior/posterior, and vermis/hemisphere are generally used in clinical studies, but these terms are anatomically vague. In functional imaging studies using positron emission tomography (PET) and fMRI when the locations of cerebellar activations are presented as Talairach coordinates (Talairach and Tournoux, 1988), it has generally not been possible to identify precisely which cerebellar lobule or folium is involved. The Talairach atlas itself has no anatomic structures depicted within the sketched outlines of the cerebellum.

Available cerebellar atlases depict major landmarks, but they are generally limited to gross morphologic relationships and they cannot localize brain structures activated in PET and fMRI studies (e.g., Crosby *et al.,* 1962; Carpenter, 1976; DeArmond *et al.,* 1976; Waddington, 1984; Roberts *et al.,* 1987; Kretshmann and Weinrich, 1992). In these atlases the individual cerebellar lobules are generally not labeled, and only limited sections are available in either one or two of the cardinal planes. More recent attempts to address these concerns using MRI (Courchesne *et al.,* 1989; Press *et al.,* 1989,1990) are useful, but they are limited by the relatively small number of sections presented and the spatial resolution of the earlier scanners.

The most detailed available atlas of the human cerebellum is that of Angevine, Mancall, and Yakovlev (1961). This dissection in three planes reveals the structure of the folia and deep nuclei. The Angevine atlas was prepared in response to a specific need, namely, to compare the cerebella of normal individuals with those of patients with alcoholic cerebellar degeneration. There are inherent difficulties, however, in using this atlas for contemporary purposes. It is applicable only in the parasagittal plane, as the coronal and axial planes of sections shown are dissimilar from those used in anatomic and functional neuroimaging studies. There are a limited number of sections in each of the cardinal planes, and there are considerable gaps between the sections. A further limiting factor of the Angevine and other atlases is that the cerebellum is not viewed within the widely used proportional stereotaxic space of Talairach and Tournoux. It is therefore not possible to determine the anatomic correlate of a Talairach coordinate derived from a functional imaging experiment. Even coregistration of a PET image with an MRI template of the cerebellum may not produce precise structure–function correlations because available atlases generally do not allow for accurate identification of the activated anatomic structure.

An additional consideration that limits the usefulness of available atlases is that the terminology used to identify the fissures and lobules is not uniform and is often contradictory. There have been many revisions of the nomenclature of the cerebellum of both human and monkey through the years (Malacarne, 1776; Vicq-d'Azyr, 1786; Meckel, 1838; His, 1895; Stroud, 1895; Flatau and Jacobsohn, 1899; Dejerine, 1901; Smith, 1902; Bradley, 1904; Bolk, 1906; Edinger, 1909; Ingvar, 1918, 1928; Langelaan, 1919; Jakob, 1928; Riley, 1929, 1930; Anatomical Society of Great Britain and Ireland, 1933; Ziehen, 1934; Ariëns Kappers *et al.,* 1936; Dow, 1942; Loyning and Jansen, 1955; International Anatomical Nomenclature Committee, 1956; Jansen and Brodal, 1958; Larsell, 1934, 1937, 1947, 1953, 1958, 1967, 1970; Larsell and Jansen, 1972). The number of attempts to understand the gross morphologic organization of the cerebellum reflects the difficulty of the task. There is considerable variation and conflict within the terminologies previously adopted. The activation seen in functional imaging studies and the precision with which MRI can determine the site of a lesion in the human cerebellum make it desirable now that each region be identified by a uniform nomenclature. It is also important to correlate hemispheric and vermal components of each lobule and to compare human cerebellar morphology with that of lower species (Larsell, 1970; Madigan and Carpenter, 1971).

This atlas was developed specifically for use in conjunction with anatomic and functional neuroimaging. It is presented in the three cardinal planes in a universally understood system of proportional stereotaxic space with sections shown at 2-mm intervals, displaying high resolution, and utilizing a revised and simplified nomenclature that avoids the terminological confusion that characterized earlier attempts.

As the project to develop this atlas evolved, it became apparent that the inability to visualize the cerebellar nuclei on the MRI sections was an important limitation that needed to be addressed. The collaboration with the Laboratory for Neuroimaging of Arthur Toga at UCLA added considerably to this effort, and made it possible to locate the nuclei on the MRI sections with a reasonable degree of certainty given that the cryosections are also presented in the Talairach plane. In labeling the nuclei on the cryosections, however, it further became apparent that it was difficult to define their boundaries. This was particularly true for the midline nuclei. In order to address this problem, histologic sections stained for Nissl substance and for myelin in the three cardinal planes were obtained and the nuclear boundaries were studied. The histologic sections were included in the atlas so that the nuclei identified on the cryosections and compared with the MR images in Talairach space could be further differentiated from each other.

We do not expect this to be the final word in atlases of MRI or gross human cerebellar morphology. The MRI and cryosection images are each derived from a single brain. Future efforts will need to address the degree to which the location of cerebellar fissures, lobules, and folia vary between individuals and across different populations. They also must determine the extent, if any, to which morphologic variability of particular lobules, folia, or nuclei correlates with behavioral measures. Further, they will need to determine whether variability of location using the Talairach coordinate system is so great as to render this system unsuitable for the cerebellum. This atlas may help facilitate in-depth and large scale analyses of structure–function correlations in the human cerebellum. It has become apparent that the cerebrocerebellar circuits are complex and highly patterned; that the clinical manifestations of cerebellar lesions include defined higher order behavioral syndromes; and that the functional imaging characteristics of cerebellum include a wide array of cognitive, sensory, and other nonmotor functions (see Schmahmann, 1997). It is our hope that the availability of this atlas will contribute to this deepening understanding of the role of the cerebellum in nervous system function.

Jeremy D. Schmahmann, M.D.

Acknowledgments

This atlas is the result of a collaborative effort by investigators in four institutions in the United States and Canada. The authors gratefully acknowledge the contributions of a number of individuals to the development of this atlas. David McDonald, Ph.D. (Montreal Neurologic Institute) developed the Display software used to locate the cerebellar fissures simultaneously in three planes, and he orchestrated the computer technology that generated the smooth fissures on the MRI sections, the three-dimensional reconstructions, and the final placement of the sections within the Talairach coordinate frame. Colin Holmes, Ph.D. (Montreal Neurological Institute, now in the Laboratory of Neuroimaging at UCLA) developed the method for averaging the 27 scans to produce a final product of great quality and clarity. Karyne Lavoie, M.S. (Laval University, Quebec City) assisted in defining the cerebellar fissures. Amy Hurwitz, M.S. (Massachusetts General Hospital) labeled the MRI images and provided invaluable assistance throughout the project. Noor Kabani, Ph.D. (Montreal Neurological Institute) helped get the project started by spending many hours acclimatizing us (J. D. S., J. D.) to the Montreal Neurological Institute brain database and the use of Display. Nikos Makris, M.D., Ph.D. and David Kennedy, Ph.D. (Center for Morphometric Analysis, Massachusetts General Hospital) provided crucial assistance at the outset of this project with respect to the possibility of cerebellar segmentation using MRI. Marygrace Neal, M.Ed. (Massachusetts General Hospital) has provided ongoing support and assistance during the completion of this work and the preparation of the manuscript. The inspired insight of Verne S. Caviness M.D., Ph.D. (Massachusetts General Hospital) emphasized the need to include the deep nuclei into this comprehensive evaluation, which had a significant impact on the project. We are grateful to Thomas L. Kemper, M.D., who provided the human brain from the Yakovlev collection from which the histologic sections in the transverse plane were derived. Deepak N. Pandya, M.D. provided the kodachromes of the brain cut in sagittal section and stained with myelin also from the Yakovlev collection, as well as the coronal sections of a brain that we are using for another study, and his invaluable assistance with the histologic analysis was also greatly appreciated. Librarians in the Rare Book Collections of the New York Academy of Medicine and in the Countway Library of Medicine at Harvard Medical School provided assistance in pursuing texts from the early and mid-eighteenth century. This work was supported in part by the National Institutes of Health through a Small Book Award from the National Library of Medicine to JDS. We are grateful to Academic Press, and particularly to Jasna Markovac, Ph.D., our editor, for her encouragement, counsel, and enthusiastic support throughout this project, and to Marge Lorang, Editorial Assistant, and Debby Bicher, Senior Graphics Coordinator, for their valuable contributions. Finally, we were particularly pleased that Raymond D. Adams, M.D. kindly agreed to write the foreword to this atlas. Dr. Adams wrote the foreword to the atlas of Angevine, Mancall, and Yakovlev in 1961 that emanated from Harvard Medical School, and we hope that this work builds sufficiently on the careful labors of our predecessors.

Jeremy D. Schmahmann, Julien Doyon,
Arthur W. Toga, Michael Petrides, Alan C. Evans

INTRODUCTORY TEXT

METHODS

Magnetic Resonance Imaging

To produce these images, signal-enhanced magnetic resonance (MR) scans were used to permit the identification of brain structures in as noise-free data as possible. A series of 27 T1-weighted MR scans were obtained from a single subject (a healthy 27-year-old male) using a Phillips 1.5 Tesla MR. The scans were acquired over a period of 3 months, with the maximum duration of any scanning session being 3 hr. A series of 20 T1-weighted scans were acquired in which the scan parameters were a 3-D sagittal volume composed of 140 1 mm^3 slices, FOV 256 mm (SI) × 204 mm (AP), acquired with a spoiled GRASS sequence (TR/TE = 18/10 ms, flip angle 30 degrees, NSA [NEX] 1). A series of seven T1-weighted scans were acquired from the same subject with the following protocol: 3-D sagittal volume composed of 200 0.78 mm^3 slices, FOV 200 mm, acquired using a T1-weighted spoiled GRASS (TR/TE = 20/12 ms, flip angle 40 degrees, NSA (NEX) 1, 6 echoes). Flow compensation was on for a total scan time of 10 hr, 50 min during the 27 scans. Each scan was registered and transformed into a standard proportional stereotaxic space, the Montreal Neurologic Institute (MNI) 305 brain-averaged space (Talairach and Tournoux, 1988), using an automatic registration tool (Collins et al., 1994) and resampled to 0.5 mm^3 before normalization and intensity averaging. The primary effect of the intensity averaging of the 27 single-image volumes was the improvement in the signal-to-noise ratio by a factor of approximately 5.2 (the expected improvement was $\sqrt{n}$, that is, a factor of 2.6 for the seven 0.78 mm^3 volumes, and 4.5 for the 20 1.0 mm^3 volumes). This effect was further accentuated by the reduction of partial volume effects. Due to the unavoidable small motions of the head between scans, each region of space was imaged in slightly different voxels in each scan. Because the final point sample was drawn from many volumes, all slightly displaced with respect to one another, the final subsampled voxels reduced the overall partial volume effect. Thus the averaging of 27 individual T1 scans resulted in an averaged volume that contained intensity data with significantly reduced noise and accentuated fine detail compared with the individual brain scans, with the net effect being substantially enhanced quality of the final MR images. The methods used in this process are further discussed in Holmes et al. (1998).

Identification of Cerebellar Fissures and Lobules

The brain slices were viewed using in-house software of the McConnell Brain Imaging Center running on Silicon Graphics workstations (Figure 1). The interactive portion of the software, Display, provided several simultaneous views of imaging data, including sagittal, coronal, and horizontal cross sections. Within each view, a variety of standard color coding algorithms was invoked to assign colors to the numerical values of the images; each view could be independently magnified and translated. The current positions of the cross-sectional planes were chosen by selecting a point on any view that caused the positions of the other views to change accordingly. This permitted cross-correlation of a given locus between the different planes. Individual voxels in the image volume were labeled with an arbitrary integer label by using the mouse to sweep out a circular brush across an image plane. In addition, a three-dimensional view of reconstructed surface geometry was available. The boundaries in an image and labeling of an image were created by an isosurface triangulation algorithm (Lorenson and Cline, 1987) and displayed in the three-dimensional window.

The coregistration of the images in the sagittal, coronal, and horizontal planes was an essential tool in the definition of both fissures and folia and facilitated the characterization of these structures with confidence. The atlas of Angevine et al. (1961) was used to identify fissures that could reliably be seen on the mid-sagittal plane. The cerebellar fissures were then identified by labeling empty voxels in the center of each fissure on every 0.5-mm-thick slice in all three planes, and this information was used to parcellate the lobules and folia. In the atlas, the sections are presented in each plane at 2-mm intervals. For those fissures that were optimally seen on more lateral sections (e.g., the superior posterior fissure), comparison was made with corresponding parasagittal sections in the atlas of Angevine et al. and validated using the coronal and horizontal images of this atlas. Eleven major fissures were identified, and these were used to subdivide the cerebellum into lobules. The differentiation of folia and subfolia was clear in most instances. The cerebellar cortex could usually be distinguished from the medullary core and medullary rays extending into the larger folia. The voxellated fissures were subjected to a least-squares curve-fitting smoothing algorithm, which assisted in the final stage of presentation of the data. The images were rendered with in-house ray-tracing software with appropriate ruled indicators delineating positions in Talairach space. Finally, the lobules were labeled according to our revision of the cerebellar nomenclature that respects the vermal–hemispheric correlation outlined in Table 1 and explained further below.

Three-Dimensional Reconstructions

The fissures marked on the MRI serial sections were transformed into smooth sheets coursing through three-dimensional space with the cerebellar tissue removed, using image blurring and the isosurface algorithm. Each set of labeled voxels was first blurred with a 5-mm-wide box filter, then the isosurface algorithm was applied to the blurred image (Lorenson and Cline, 1987). The resulting triangulated surface constituted a visually smooth representation of the volumes of the fissures. These fissures suspended in the invisible cerebellum are shown in Figure 2. The external surface of

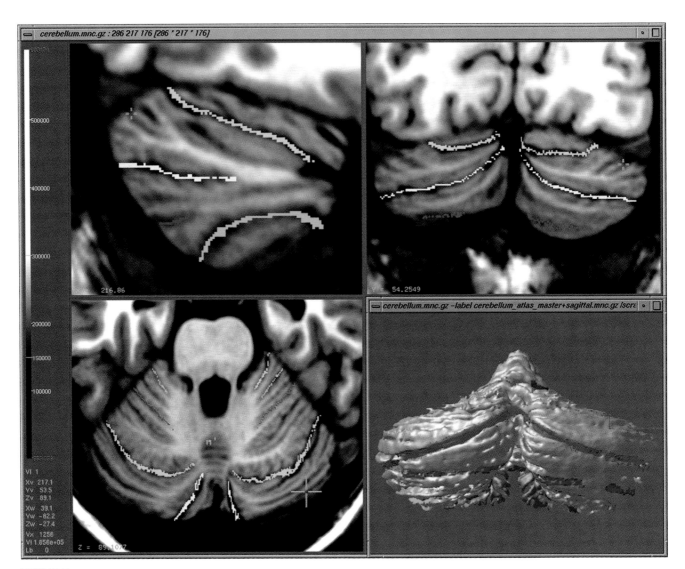

FIGURE 1

View of the Display software used in this study. The cursor is visible on the sagittal (top left), coronal (top right), and horizontal (bottom left) views of the cerebellum, as well as on a 3-D reconstruction of the surface (bottom right).

TABLE 1
Relationship of the Fissures to the Cerebellar Lobules in the Vermis and Hemispheres as Defined By and Used in This Atlas

VERMIS Lobule	FISSURE	HEMISPHERE Lobule
I,II		I,II
	Precentral	
III		III
	Preculminate	
IV		IV
	Intraculminate	
V		V
	Primary	
VI		VI
	Superior Posterior	
VIIAf		Crus I
	Horizontal	
VIIAt		Crus II
	Ansoparamedian	
VIIB		VIIB
	Prepyramidal/Prebiventer	
VIIIA		VIIIA
	Intrabiventer	
VIIIB		VIIIB
	Secondary	
IX		IX
	Posterolateral	
X		X

the cerebellum was reconstructed by applying the isosurface algorithm to the average MR image. The surface markings of the fissures were identified on these views, and the cerebellum thus parcellated into lobules (Figure 3). These three-dimensional reconstructions of the surface of the MRI cerebellum with the superimposed externally visible fissures provided independent confirmation of the accuracy of the labeling of the fissures.

Cryosection Images

The MRI investigation was not specifically designed to study the detailed anatomy of the deep cerebellar nuclei. The outline of the nuclei characterized by the gray-white matter differentiation proved indistinct on these images. It was therefore difficult to define their limits with certainty. This was true for the nuclei as a group, as well as for the delineation of the fastigial, globose, emboliform, and dentate nuclei from each other. This limitation was compensated for by comparison with postmortem cryosections of another cerebellum.

A spatially accurate, high-resolution 3-D volume of brain anatomy was obtained from a cryosectioned whole human head from the cadaver of an adult female. The cryosectioning was performed using a heavy-duty cryomacrotome (PMV, Stockholm, Sweden) modified for quantitative digital image capture. Serial images (1024(2), 24-bit) were captured directly from the cryoplaned specimen blockface in 200-micron intervals and reconstructed to a 3-D data volume. Data were placed into the Talairach coordinate system to create a volume of brain anatomy for atlas reference. This volume was resampled at 500 microns along the sagittal, coronal, and horizontal planes and the images were enhanced by digitally editing the background. The spatial resolution of the cryosection images was 170 microns/pixel for the whole and 40 microns/pixel for isolated brain regions. Anatomic detail was superior to the MRI, particularly for deep structures such as the deep cerebellar nuclei. The digital repositioning in the Talairach coordinate system enabled efficient structure localization and morphometric comparison. Further details of this methodology are described by Toga *et al.* (1994, 1997).

In this atlas, equivalent sections of the cerebellum in the cardinal planes were selected for comparison with the MRI cerebellum. This was performed by computing a linear transformation. In each of the three planes, adjacent cryosections sampled at 500-micron intervals were available for analysis. Those sections that included the cerebellum were identified. For each of the three planes, the number of available 0.5-mm-thick cryosections was divided by the number of 2-mm interval 0.5-mm-thick MRI images presented in this atlas. The resulting integer represented the interval between cryosections required to match the MRI sections. In using this linear transformation method, we recognize that the MRI and cryosections are not identically matched, but the likelihood for error was minimized by the fact that both

data sets were placed in the Talairach coordinate system. The correspondence between the planes in the MRI brain and in the cryosection brain was excellent for the sagittal plane, good for the coronal plane, and slightly off for the transverse or horizontal plane because of "wobble" or slight misalignment of the cerebellum with respect to the cerebral hemispheres. For the most part, however, the correspondence between the two data sets was reasonably precise, and the anatomic variability evident in the demarcation of the fissures and lobules is likely to reflect true anatomic variability between individual brains rather than sampling error in the methodology.

The deep cerebellar nuclei were readily located and identified on the cryosections by comparisons with published anatomic atlases. Further, their location and configuration could be used to predict their location within the cerebellar parenchyma on the corresponding MRI images. The labels applied to the fissures, lobules, and folia in the cryosections were derived from a comparison with the MRI data and are therefore not primary data as they are for the MRI sections. This translation of the labels for the fissure and lobules from the MRI to the cryosections demonstrates the utility of the MRI atlas, even with the Talairach reference grid removed. That is, the anatomic features that define the cerebellar lobular organization in the MRI brain appear to be consistent enough that similar landmarks could be identified in the cryosectioned brain.

Histological Analysis of the Deep Cerebellar Nuclei

Unlike the MRI or the cryosection brains, the histologic images of the cerebellar nuclei were obtained from 3 separate individuals.

The myelin stained sagittal sections were derived from a set of teaching kodachromes of a brain in the Yakovlev collection that is now housed at the Armed Forces Institute of Pathology in Washington, D.C. The nuclei are adequately visualized on these sections, although adjacent Nissl stained sections were not available for comparison. Sections were chosen that corresponded with sections from the cryosection brain, although this correlation was necessarily approximate. The right side only is shown as this demonstrates the nuclei, and is relevant for the conclusions reached about the nuclear boundaries in the sagittal cryosections in both the right and left hemispheres.

The coronal sections were derived from the brain of an adult woman who had no neurological disease. The brain was immersed in formalin for four months prior to being embedded in celloidin. It was then sectioned at 35-micron intervals in the coronal plane, stained for Nissl substance with cresyl violet and for myelin using a Loyez stain, and then mounted between glass slides. The plane of section of the histologic brain is roughly similar to that of the cryosection brain, and this is supported by comparison of the gross

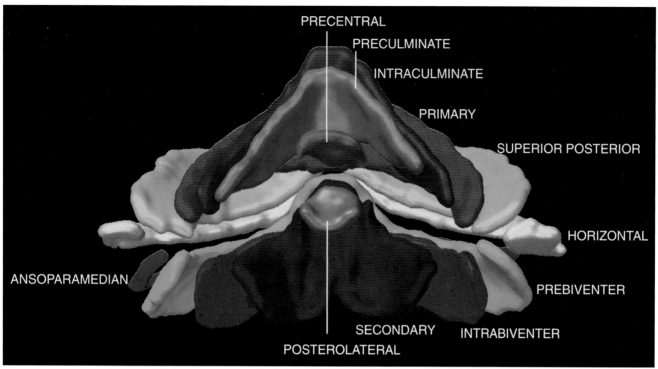

A

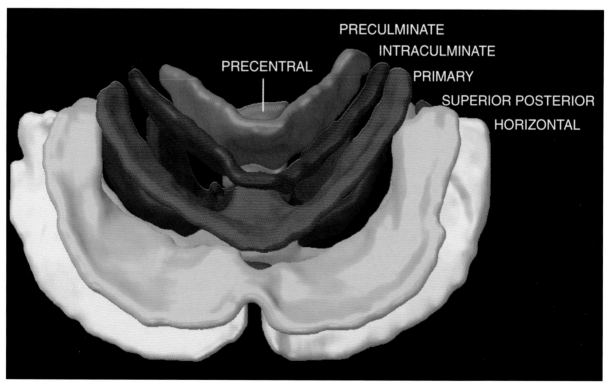

B

FIGURE 2

Three-dimensional reconstructions of the named fissures with cerebellar tissue removed to demonstrate the relationship of the fissures to each other. (A) Anterior view, (B) superior view, (C) right lateral view, (D) left lateral view.

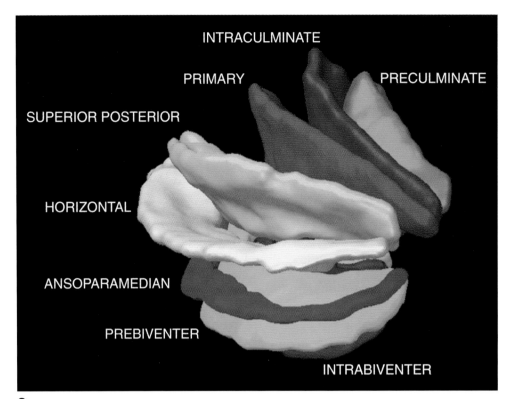

C

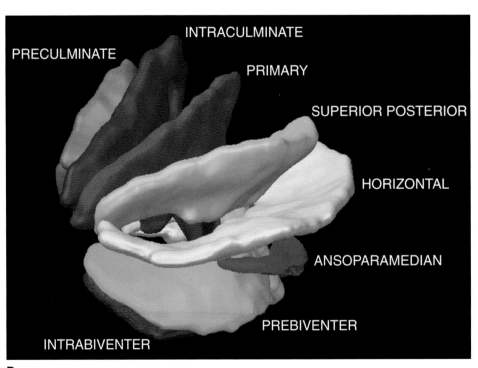

D

FIGURE 2 *(Continued)*

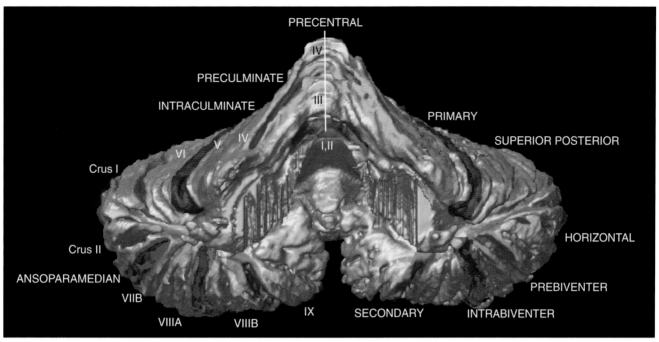

A

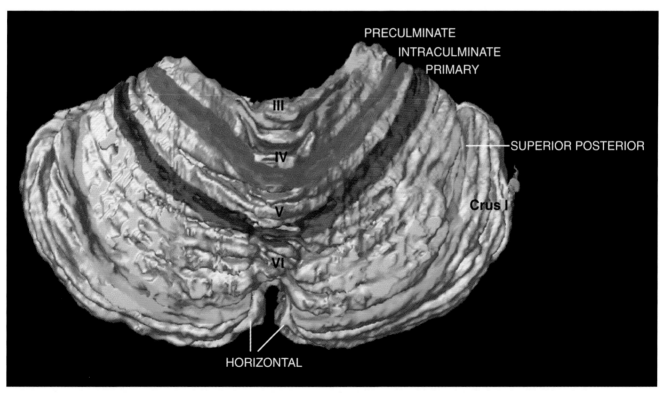

B

FIGURE 3

Three-dimensional reconstructions of the external surfaces of the cerebellum with the fissures demarcated in color. The cerebellum is viewed from the (A) anterior, (B) superior, (C) posterior, (D) inferior, (E) right lateral, and (F) left lateral aspects.

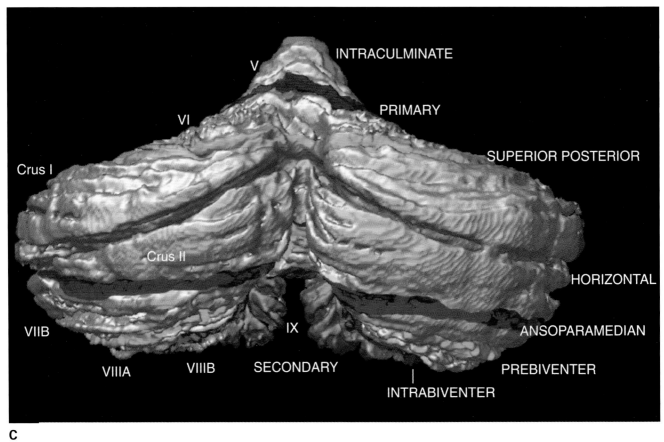

C

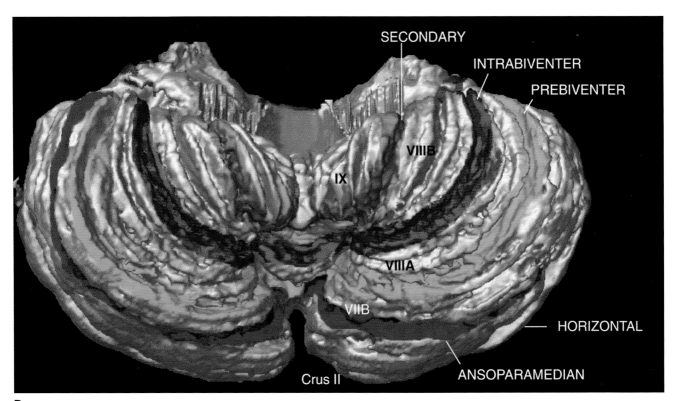

D

FIGURE 3 *(Continued)*

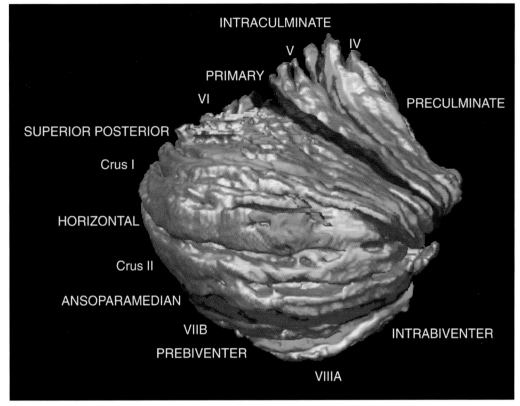

E

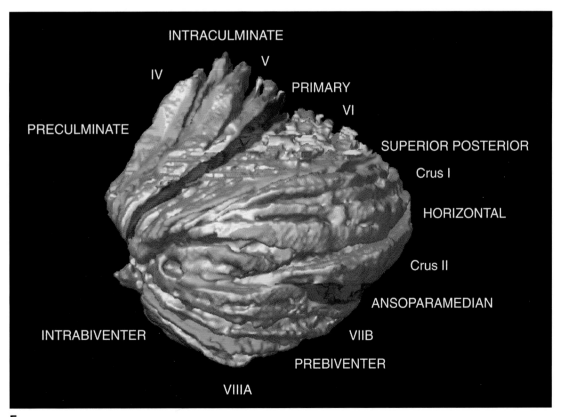

F

FIGURE 3 *(Continued)*

sections of the histology brain with Figure A6 of the Appendix of Damasio and Damasio (1988) and with the figures in the Talairach atlas referred to above. This is further supported by comparison of the nuclei as a group in the histology and the cryosection brains.

The horizontal sections were derived from the brain of a 75-year-old woman that was also originally part of the Yakovlev collection when it was still housed at the Neurological Unit of the Boston City Hospital. The brain was embedded in celloidin, sectioned in the horizontal plane at 35 microns, stained for Nissl substance with cresyl violet and for myelin using a Loyez stain, and then mounted between glass slides. The axis of these horizontal sections does not correspond precisely with that of the Talairach atlas. It is not possible to reconstruct this precisely, but by comparing the gross anatomy of the sections with those of Figure 4 of the Appendix of Damasio and Damasio (1989) and with those of Figures 2, 10, 12, 14, and 16 of Talairach and Tournoux (1988), it appears that the plane of section is between 15 to 20 degrees tilted (anterior upwards) off the anterior commissure–posterior commissure (ac–pc) line. This is apparent when attempting to compare the nuclei as a group on the horizontal histologic sections with those of the cryosections. This is exemplified by the fact that on the histology sections more of the denate nucleus (that lies more caudally) is visible when the midline nuclei (that lie more rostrally) are prominent than is the case with the cryosections (see $z=-27$ through $z=-31$).

The determination of the nuclear boundaries was performed in consultation with the atlases and descriptions of Ziehen (1934), Jansen and Brodal (1958) as discussed and illustrated in Larsell and Jansen (1972), Riley (1960), Angevine *et al.* (1961), Miller and Burack (1977), Fix and Punte (1981), Voogd *et al.* (1990), and *Gray's Anatomy 38th Edition* (1995), and by histologic analysis of the nuclei on the available sections. The cerebellar nucleus termed the basal interstitial nucleus identified in the monkey atlas of Paxinos *et al.* (2000) that lies anterior and lateral to the main body of the nuclear group and immediately adjacent to the roof of the fourth ventricle appears to have a homologous region in the human cerebellum as seen at higher power ($10\times$) in both the myelin and the Nissl stained histologic sections, but this nucleus was not specifically labeled in the present work.

Methodological Considerations

Single Subject

There are important limitations that arise from the use of a single brain to develop a cerebellar atlas. A number of variables are by definition excluded from consideration. It is not possible, for example, to evaluate consistency of hemispheric asymmetry and the individual variation of the size, shape, and constituent elements of an individual lobule. Some possible confounding variables such as gender, age, and hand-

edness are also not factored into this evaluation. The utility of the approach used in this in-depth analysis of a single brain, however, is that the primary step is the identification of the major fissures, and the determination of the lobules and folia follows from this first step. The anatomic location of major fissures can be ascertained in other cerebella by reference to the anatomic features of this atlas, even if there is some variability in their location within standard stereotaxic space and some divergence from the exact shape of this particular cerebellum. This assertion is supported by the labeling of the cryosectioned brain, based on the data from the MRI brain. This atlas can be used as the template for future population-based morphometric and probabilistic studies that examine large numbers of cerebella and that can address questions of individual variability determined by, or concomitant with, such considerations as gender, handedness, intelligence, dexterity, and other functions now seriously considered to implicate cerebellum (see Schmahmann, 1997).

Talairach Coordinate Space

It could be argued that the use of Talairach proportional space is suboptimal for an atlas of the cerebellum because the anterior–posterior commissure line is far removed from the posterior fossa. The concern, therefore, is that the margin of error in reporting the location of structures (or structure–function correlations) in the cerebellum is unacceptably great. This concern may be valid, and the degree to which the Talairach system remains viable in reporting anatomic landmarks or functional activations in the cerebellum will be determined by future studies. At this time, however, the Talairach system is widely used, and activations on functional imaging scans are commonly reported in both cerebral and cerebellar structures on the same sagittal/coronal/horizontal images. It seems possible to develop a system of coordinates specifically for use with the posterior fossa, but the overall utility of this approach may be less generally applicable. The degree to which the locations of the cerebellar folia are consistent from one brain to the next using this stereotaxic proportional space will need to be determined in future investigations. This concern does not detract, however, from the validity of the gross morphologic identification in this cerebellar atlas that was made possible by the use of contemporary methodology.

Cerebellar Nuclei

The absence of clearly visible boundaries of the cerebellar nuclei on the MRI prevented the confident incorporation of the nuclei into the MRI sections of this atlas. This limitation was addressed by comparison with the digitized autopsy cryosections of cerebellum warped into Talairach space. Because of expected individual variability between the two brains, the decision was made not to directly superimpose the MRI and cryosectioned cerebella; rather, sections at equivalent levels are presented by using a linear transformation

TABLE 2A

Comparison of Earlier Nomenclature Systems with the Designation Used in This Atlas for (A) the Vermis and (B) the Hemispheres[a]

Schmahmann et al. (1999)	Consensus	Dejerine (1901)	Bolk (1906)	Ingvar (1918, 1928)	Riley (1929)	Larsell (1936-1972)
I,II	lingula	lingula	1	lingula	lobulus I vermalis	I, II lingula
III	centralis	central	2	centralis	lobulus II vermalis	III centralis
IV	culmen (culmen of monticulus; culminis)	culmen	3	culmen	lobulus IV vermalis	IV culmen
V			4			V culmen
VI	declive (declive of monticulus; clivus monticuli; lobulus clivi)	déclive	lobulus simplex	simplex	lobulus C2 vermalis, pre-sulcal	VI declive
VIIAf	folium (folium of vermis; folium cacuminis)	bourgeon terminal	C₂	lobulus medius	lobulus C2 vermalis, post-sulcal	VIIA folium/tuber
VIIAt	tuber (tuber of vermis; tuber valvulae)	tubercule valvulaire		medianus		
VIIB	caudal aspect of tuber valvulae					VIIB caudal aspect of tuber
VIIIA	pyramis	pyramide	C₁	pyramis	lobulus C1 vermalis	VIIIA pyramis
VIIIB						VIIIB pyramis
IX	uvula	luette	b	uvula	lobulus B vermalis	IX uvula
X	nodulus	nodule	a	nodulus	lobulus A vermalis	X nodulus

[a]Based on Larsell.

TABLE 2B

Comparison of Earlier Nomenclature Systems with the Designation Used in This Atlas for (A) the Vermis and (B) the Hemispheres[a]

Schmahmann et al. (1999)	Henle (1879)	Schwalbe (1881)	Kuithan (1895)	BNA (1895)	Flateau & Jacobson (1899)	Dejerine (1901)	Smith (1902)	Bolk (1906)	Schäfer & Symington (1908)	Ingvar (1918, 1928)
I,II	vinculum lingulae	frenulum lingulae	lingula	vinculum lingulae cerebelli	lingula	frein de la lingula	lingula	1	fraenulum lingulae	lingula
III	lobus centralis	lobus centralis	ala lobuli centralis	ala lobuli centralis	ala lobuli centralis	aile du lobule central	pars preculminata	2	ala lobuli centralis	centralis
IV	lobus quadrangularis	lobus lunatus anterior	lobus lunatus anterior	lobulus quadrangularis, pars anterior	lobulus quadrangularis	partie antérieure du lobe quadrilatére	pars culminus (pars culminata)	3	lobulus lunatus anterior	culmen
V								4		
VI		lobus lunatus posterior	lobus lunatus posterior	lobulus quadrangularis, pars posterior		partie postérieure du lobe quadrilatére	area lunata	lobulus simplex	lobulus lunatus posterior	lobulus simplex
Crus I	lobus semilunaris superior	lobus posterior superior	lobus semilunaris superior	lobulus semilunaris superior	lobulus semilunaris superior	lobe semilunaire supérior	area pteroidea	lobulus ansiformis, crus I	lobulus posterior superior	lobulus ansiformis
Crus II	lobus semilunaris inferior	lobus posterior inferior	lobus semilunaris inferior	lobulus semilunaris inferior	lobulus semilunaris inferior	lobe semilunaire inférieur	area post-pteroidea	lobulus ansiformis, crus II	lobulus semilunaris inferior	
VIIB				lobulus gracilis		lobe grêle			lobulus gracilis	
VIIIA	biventer	lobus cuneiformis sive biventer	lobus cuneiformis	lobulus biventer	lobulus biventer	lobe digastrique	area parapyramidalis	lobulus paramedianus	lobulus biventralis	
VIIIB										
IX	tonsilla	tonsil	tonsil	tonsil	tonsil	amygdale		crus circumcludens	tonsil	lobulus paramedianus/tonsil
X	flocculus	flocculus	nodulus	flocculus	flocculus	flocculus	flocculus	pars floccularis	flocculus	flocculus

[a]Based on Larsell.

Langelaan (1919)	Riley (1929)	BNA-BR (1933)	Ziehen (1934)	Ariëns Kappers et al. (1936)	Dow (1942)	NAP (1955)	Jansen and Brodal (1958)	Larsell (1936–1972)	Angevine et al. (1961)	Press et al. (1989)
lobus vinculo-lingualis	lobulus I lateralis	lingula	vinculum lingulae	lingula	lingula	lingula cerebelli	lobulus I	H II lingula	lingula	lingula
lobus centro-alaris	lobulus II lateralis	ala lobuli centralis	ala lobuli centralis	alal lobuli centralis	lobulus centralis	ala lobuli centralis	lobulus centralis, II, III	H III centralis	lobulus centralis	central lobule
lobus culmino-lunatus	lobulus IV lateralis	lobulus lunatus anterior	lobus quadrangularis, pars anterior	lobus quadrilaterus anterior	culmen	lobus quadrangularis	lobulus quadrangularis anterior	H IV culmen H V	culmen	quadrangular lobule, anterior portion
lobus declivo-lunatus	lobulus ansiformis, crus 1	lobulus lunatus posterior	lobus quadrangularis, pars posterior	lobus quadrilaterus posterior	lobulus simplex	lobulus simplex	lobulus quadrangularis posterior	H VI lobulus simplex	lobus simplex	quadrangular lobule, posterior portion
lobus folio-semi-lunaris		lobulus ansiformis, facies superior	lobulus semi-lunaris superior	lobus semi-lunaris superior	crus I of lobulus ansiformis	lobulus semi-lunaris superior	lobulus semi-lunaris superior	crus I of H VIIA/ lobuli ansiformis	crus I, lobulus ansiformis	semilunar lobule, superior portion
lobus tubero-semi-lunaris	lobulus ansiformis, crus 2	lobulus ansiformis, facies inferior	lobulus semi-lunaris inferior	lobus semi-lunaris inferior		lobulus semi-lunaris inferior	lobulus semi-lunaris inferior	crus II of H VIIA/ lobuli ansiformis	crus II, lobulus ansiformis	semilunar lobule, inferior portion
				gracilis	crus II lobulus ansiformis		lobulus gracilis	H VIIB lobulus paramedianus	lobulus paramedianus	gracile lobule
lobus pyramido-biventricus	lobulus para-medianus		lobulus biventer (2 parts)	crus secundum lobi ansiformis		lobulus biventer	lobulus biventer/paraflocculus dorsalis	H VIIIA lobulus biventer, pars copularis	lobulus biventer, pars copularis	biventer
								H VIIIB l. biventer, pars paraflocculus dorsalis	lobulus biventer, pars paraflocularis dorsalis	
lobus uvulo-tonsillaris	lobulus para-floccularis	tonsilla	tonsilla	lobus para-medianus (tonsilla)	lobulus para-medianus	tonsilla	tonsilla/paraflocculus ventralis	paraflocculus ventralis	paraflocculus	tonsil
lobus nodulo-floccularis	lobulus floccularis	flocculus	flocculus	flocculus	flocculus	flocculus	flocculus	flocculus	flocculus	flocculus

paradigm, as discussed previously. The nuclei were identified on the cryosections by comparison with available atlases, and by reference to the histologic sections embedded in celloidin and stained for Nissl substance and myelin as presented in the atlas. The histologic sections in the sagittal and coronal planes correspond reasonably closely with the nuclei in the cryosection brain, but the horizontal planes of section are approximately between 15 to 20 degrees out of alignment. This does not negate the principal reason for including the nuclear histology, namely enhancing the ability to differentiate between the individual nuclei, but it does lessen the power of the comparison between the MR images, the cryosection brain, and the histologic sections.

Nomenclature

What began as a straightforward project of mapping the fissures and the nomenclature used by Angevine, Mancall, and Yakovlev (1961) onto the MRI images turned out to be more complicated. The nomenclature systems used by both early and contemporary investigators are confusing. We attempted to resolve this issue by reviewing the evidence in support of the different terminologies presented by earlier investigators, starting with that of Vincenzo Malacarne in 1776. The atlas of Angevine *et al.* also helped us revise the nomenclature and determine the relationship between the fissures and lobules, because their atlas included an analysis, comparison, and synthesis of the findings of many of the earlier investigators. By using the current technology (MRI reconstructions and Display software that permits simultaneous cross-referencing between the three planes) we were able to explain the origins of some terminological difficulties that have arisen over the past two centuries and then resolve them. We amended the terminology of Larsell, and compared this with the terminologies introduced by Malacarne (1776), Bolk (1906), Ziehen (1934), and others. In so doing we adopted a nomenclature that is anatomically correct, is easily understandable, and melds with the neuronames brain hierarchy (Bowden and Martin, 1995) accepted by the International Consortium for Brain Mapping.

The method employed for determining the simplified nomenclature was derived from the use of the atlas itself. Vermal lobules were identified with the assistance of the midsagittal section of the Angevine atlas, subsequent to the identification of the fissures and sulci. The lobules that could be readily discerned in the hemisphere were also demarcated, such as crus I and crus II, which are separated by the horizontal fissure that is prominent in lateral sections. By applying a slightly modified version of the Larsell (1970; Larsell and Jansen, 1972) terminology and maintaining consistency through all planes in continuous adjacent 0.5-mm sections, it was possible to determine which folium belonged to which lobule, and which Larsell designation was appropriate. The descriptions and illustrations of the rhesus

monkey cerebellum by Larsell (1970) and Madigan and Carpenter (1971) were also helpful in furthering our understanding of the folia of the human cerebellum. Some of the small subfolia buried at the fundus of the fissures or sulci were troublesome, but most could be parcellated by use of the three-dimensional correlations.

The names of the fissures were not changed, because they are anatomically consistent, have historic significance, and are still widely used and understood. In contrast, we dispensed with the Latin names of the lobules and used Larsell's Roman numeral designations with some revisions. Table 1 presents the nomenclature used in this atlas to describe the cerebellar fissures and lobules. Table 2 presents a comparison of the earlier nomenclature systems with that derived from this atlas (based on Larsell). This table is helpful when drawing anatomic correlations between studies that employ different nomenclatures.

The Problem of the Vermis

There is no true "vermis" in the anterior lobe. The application of this term to the midline and paramedian sectors of the anterior lobe is an extension of the Latin term (meaning worm) used by Malacarne to denote the structure visible in the posterior and inferior aspects of the cerebellum. The vermis (as such) is present from lobules VI through X. The use of the term vermis to indicate "midline" has become well entrenched, however, and so it has brought with it the problem of defining what is the lateral extent of the anterior lobe "vermis." It has been suggested that the paravermian sulcus limits the vermis laterally. In many brains there is no paravermian sulcus, and where one appears to be present, it may simply reflect the indentation produced by the course of the medial branch of the superior cerebellar artery. On coronal section, the anterior lobe (lobules I through V) has cortex buried beneath the overlying midline cortex, separated from the hemispheral cortex by a superior white matter extension of the medullary core (see, e.g., coronal sections $y = -42$ through $y = -58$). Even where there may be a paravermian sulcus, this is most likely the preculminate fissure (see coronal section $y = -42$) or the intraculminate fissure ($y = -48$) that is not necessarily constant from one section to the next, and this does not demarcate the lateral extent of the hidden cortex that truly occupies a midline position between the hemispheres. In the MRI brain, the anterior lobe "vermis" occupies a mediolateral extent that varies from 14 to 20 mm depending on the section and the definition of the lateral boundary (surface markings of the "paramedian sulcus"; lateral extent of the buried cortex; lateral aspect of the paramedian white matter laminae). Each brain will no doubt vary to some extent, and the functional significance of this distinction is not known. The determination of the lateral extent of the "vermis proper" (in the posterior lobe) is not a problem, as this is a gross morphologic term originally intended to define a structure visible on the surface of the brain.

Larsell distinguished the vermis from the hemispheral compartments of the different lobules by using the prefix "H" for the hemispheres. Because the bulk of the cerebellum is composed of hemisphere, and the H designation adds a cumbersome element to the description of the lobules, we dispensed with this designation. Instead, when referring to a midline structure, the term vermal area (anterior lobe) or vermis (posterior lobe) was added.

RESULTS

Three-Dimensional Images

The three-dimensional reconstruction of the fissures provides a new view of the fissures as they course through the cerebellum and subdivide it into lobules (Figure 2). The three-dimensional renditions of the surface views of the cerebellum reveal the external location of the fissures and the parcellation of the cerebellar surface into lobules.

Views of the Cerebellum in the Sagittal, Coronal, and Horizontal Planes

Sections of the cerebellum are presented at 2-mm intervals in the parasagittal, coronal, and horizontal planes, showing the progression of the fissures and lobules from midline to lateral, anterior to posterior, and superior to inferior. The fissures are color coded according to the table. The conventions used in the atlas of Talairach and Tournoux (1988) are adopted here. Thus right and left sides are $+x$ and $-x$, respectively. Distance anterior and posterior to the ac–pc line is $+y$ and $-y$, respectively. Distance above and below the ac–pc line is $+z$ and $-z$, respectively. In the horizontal series the distinction between the vermal and hemispheric nomenclature on lobule VII (i.e., VIIAf-Crus I and VIIAt-Crus II) is not strictly adhered to for practical reasons, but these designations are preserved in the sagittal and coronal planes of section. Cryosections of the postmortem brain at equivalent levels show the locations and anatomic structures of the deep nuclei as well as the lobular organization.

Descriptions of the Fissures and Lobules

The descriptions of the fissures and lobules apply specifically to the MRI brain. The determination of the fissures and thus of the lobules was verified in every section in each of the three cardinal planes with reference to the other planes. The designations of the fissures and lobules in the cryosection brain were derived from the MRI brain by matching the features of equivalent sections, while at the same time main-

taining internal consistency of the identification of the fissures and lobules in the cryosections images. Whereas they appear to correspond well to the MRI brain and thus seem to be accurate, the anatomic statements for the postmortem cryosection brain should be viewed as derivative, rather than primary data.

Vermal lobules I and II were indistinguishable, and abut the anterior medullary velum that lies anterior and inferior to the vermis. There is no hemispheric extension of vermal lobule I. Hemispheric lobule II is the lateral extension of I, II. Lobule I is distinct from lobule II in some human cerebella and in lower primates. This distinction is thus maintained, even though it is not always apparent. The vermal and hemispheric components of lobule I/II have together been termed the lingula.

The precentral fissure separates lobules I and II from lobule III (centralis), both in the vermis and in the hemispheres. Lobule III is partially obscured by lobule IV when viewed from the superior aspect, but it is sizeable, has a semilunar form, and is the first distinct folium not attached to the superior medullary velum.

The preculminate fissure separates lobule III from lobule IV.

The use of the older term "culmen" (including both the vermis and the hemisphere) obscures the fact of the division of the culmen into lobules IV and V by the intraculminate fissure. This fissure is readily discernible on the hemisphere, but at the midline is continuous with a small fissure situated between the branches of a folium and thus can be difficult to localize with accuracy in the absence of adjacent sagittal sections or surface reconstructions. In this brain it was possible to trace the intraculminate fissure from the right hemisphere across the midline to the left side. By using the coronal images (see $y = -46$) we determined that the three major folia of the "culmen" are subdivided such that lobule IV contains one folium and lobule V contains two.

The primary fissure distinguishes the anterior lobe of the cerebellum (lobules I through V) from the posterior lobe (lobules VI through IX); specifically, it separates lobule V from lobule VI, both in the vermis and the hemisphere. The primary fissure and other fissures within the anterior lobe are progressively more difficult to discern on parasagittal sections as one moves laterally away from the midline. The primary fissure is unmistakable on the midsagittal section, but it is continuous with an undistinguished small fissure in the intermediate sectors of the hemispheres.

The superior posterior fissure appears early in development and is visible in this brain on the 3-D reconstruction of the superior surface (Figure 3). Lobule VI lies between the primary and superior posterior fissures and is therefore sizeable, and it contains two large folia with medullary rays that take their origin off the medullary core (e.g., see $y = -58$). The vermal and hemispheric components of lobule VI have previously received different names. Lobule VI in the vermis was called the declive, and in the hemisphere it has received the name lobulus simplex, among others (see Table 2).

The superior posterior fissure separates lobule VI from lobule VII in the vermis, and lobule VI from crus I (of the ansiform lobule) in the hemisphere.

Vermal lobule VII is complex, and its lateral extension into the ansiform lobule of Bolk forms a large part of the hemisphere in the human cerebellum. The terms folium and tuber have been used to denote vermis VII, but this fails to convey the complexity of the equivalent hemispheric extensions. Further, VIIA (in the Larsell terminology) is divided by the horizontal fissure into two sectors, with important differences in their hemispheric extensions.

Vermal lobule VIIAf (previously termed the folium) expands laterally to form crus I (of the ansiform lobule). Vermal lobule VIIAt (previously the rostral half of the tuber) corresponds in the hemisphere with crus II. The horizontal fissure separates VIIAf from VIIAt in the vermal region, and crus I from crus II in the hemispheres. In this MRI brain, VIIAf is not present in the midline. It appears 6 mm to the right of midline and 4 mm to the left of midline. For this reason, in the midsagittal plane and just lateral to it, the superior posterior fissure merges with the horizontal fissure and vermal lobule VI is bounded caudally by VIIAt. A similar arrangement is found in the cryosection brain, in that VIIAf appears 4 mm to the right of midline and 8 mm to the left of midline. In some uncut postmortem cerebella that we have examined, a very small folium may be seen in the midline separating the superior posterior from the horizontal fissures, and thus we can confirm the presence of a small VIIAf in some individuals. In the two brains in this atlas, however, VIIAf (or the folium of vermis, in the old terminology) is not continuous across the midline.

The ansoparamedian fissure is submerged on the ventral surface of the "tuber," separating lobules VIIAt from VIIB (previously termed the gracile, or paramedian lobule). The precise delineation of the ansoparamedian fissure and the division of crus II from VIIB in the hemisphere have long been the subject of controversy. We were able to delineate the ansoparamedian fissure, but in following it from the midline laterally, an asymmetry was revealed across the hemispheres. This occurred because on the left side the horizontal and ansoparamedian fissures meet on the posterior inferior cerebellar surface. Crus II on the left therefore is a triangular lobe situated on the superior and the posterior aspects of the posterior lobe. On the right side, however, the horizontal and ansoparamedian fissures do not meet, thus revealing that crus II continues laterally and forward onto the anterior surface of the cerebellum. The effect of this asymmetry is that on lateral sagittal sections, crus I is adjacent to lobule VIIB on the left, whereas on the right crus II is adjacent to VIIB. This asymmetry may help resolve some of the earlier debate as to whether VIIB (gracile, or paramedian lobule) is part of, or distinct from, the inferior semilunar lobule (see, e.g., His, 1895, Basle Nomina Anatomica;

Smith, 1902; Ziehen, 1934; Larsell and Jansen, 1972). Coronal sections in this brain (see $y = -54$ through -74) confirm that despite the fact that the folia at the lateral inferior cerebellar surface are asymmetric across the two hemispheres, they are derived from medullary rays that are morphologically similar.

The gross morphologic organization of the "tuber" and its hemispheric extensions deserves further comment. The "tuber" in the vermis is bounded rostrally by the horizontal fissure. As already noted, the horizontal fissure can be very small in the midline or merged with the superior posterior fissure. In the midsagittal section, therefore, the "folium" (VIIAf) that lies rostral to the "tuber" is small or even absent (as in the midsagittal sections of both the MRI and the cryosection brains) despite the fact that it evolves laterally into sizeable crus I (of the ansiform lobule). The "tuber" is bounded caudally in the midsagittal section by the prominent prepyramidal/prebiventer fissure, separating lobule VIIB from lobule VIIIA. Thus the bulk of the "tuber" contains VIIAt, and its undersurface (not visible from the external surface of the cerebellum, but seen only on the midsagittal section) comprises VIIB. The hemispheric extension of VIIAt is crus II (of the ansiform lobule). The hemispheric extension of vermal VIIB is still VIIB in the new nomenclature (gracile lobule or paramedian lobule in prior terminologies). This has direct bearing on the debate over the preceding two centuries about the constituents of the "inferior semilunar lobule." The hemispheric extension of the "tuber" as such (VIIAt and VIIB) is thus crus II and VIIB. In the older terminologies, then, "tuber" of vermis was continuous with whatever those two lobules were called, that is, inferior semilunar lobule including VIIB, or inferior semilunar lobule plus the gracile/paramedian lobule that we now know as VIIB. This atlas demonstrates that there is good reason to be precise in the definitions of these lobules, as the revised nomenclature of this atlas proposes. VIIA may be small in the vermis, but the ansiform lobule that it is continuous with in the hemisphere (VIIAf–crus I; VIIAt–crus II) is very large. Further, crus II and hemispheral VIIB are asymmetric across the two sides. Whether this has functional significance or not, the anatomic variability identified through the years that stems from lumping together the tuber-ansiform and gracile/paramedian lobules has been sufficiently confusing that it warrants this more precise delineation. Hence, the vermal–hemispheric equivalents are VIIAf–crus I; VIIAt–crus II; VIIB–VIIB.

The prepyramidal/prebiventer fissure separates VIIB from VIIIA. It is named according to whether the VIIB/VIIIA distinction is in the vermis (prepyramidal fissure) or hemisphere (prebiventer fissure). The intrabiventer fissure separates VIIIA from VIIIB. (The term "intrapyramidal" was omitted for purposes of simplification.)

The secondary fissure is interposed both at the vermis and in the hemispheres between VIIIB and IX (in the older terminology—uvula at the vermis; tonsil, or ventral and accessory paraflocculus, at the hemisphere).

The posterolateral fissure forms the boundary between the posterior lobe of the cerebellum and the flocculonodular lobe, separating IX from X (in the older terminology—nodulus at the vermis; flocculus at the hemisphere).

CONCLUSION

The understanding of the human cerebellum has been substantially advanced by the availability of detailed images revealed by anatomic MRI and by cerebellar activation with PET and fMRI studies of motor, sensory, and cognitive/emotional functions (see, e.g., Desmond *et al.,* 1998; Doyon, 1997; Fiez and Raichle, 1997; Parsons and Fox, 1997). Until now, determining the precise location of lesions or sites of activation within the cerebellum has been limited by the lack of a sufficiently detailed and practical atlas.

This atlas has relied upon high-resolution anatomic images using both MRI and cryosectioned postmortem material, Display software that permits simultaneous visualization of a cursor in the three cardinal planes within the cerebellum, and 3-D reconstruction of the data. The resulting atlas of the human cerebellum extends the understanding of the gross morphology and organization of the cerebellum to a previously impenetrable level. It clarifies the anatomy of the human cerebellum, provides correlations of vermal with hemispheric structures, and defines the cerebellar lobules and folia. It demonstrates the details of the cerebellar cortex within the three cardinal planes in Talairach proportional stereotaxic space and makes this material available and useful to clinicians and investigators interested in the study of the cerebellum. An immediate consequence of this new understanding of cerebellar anatomy is a revised, more contemporary and accessible cerebellar nomenclature.

REFERENCES

Anatomical Society of Great Britain and Ireland. (1933). "Final report of the committee appointed by the Anatomical Society of Great Britain and Ireland on June 22, 1928." Robert Maclehose and Co., Ltd. University Press, Glasgow. (Birmingham Revision of the Basle Nomina Anatomica; BNA-BR).

Angevine, J. B., Mancall, E. L., and Yakovlev, P. I. (1961). "The Human Cerebellum: An Atlas of Gross Topography in Serial Sections." Little, Brown and Company, Boston.

Ariëns Kappers, C. U., Huber, G. C., and Crosby, E. C. (1936). "Comparative Anatomy of the Nervous System of Vertebrates, Including Man, Vol 1." Macmillan Publishers, New York.

Bowden, D. M., and Martin, R. F. (1995). Neuronames brain hierarchy. *Neuroimage* **2,** 63–83.

Bradley, O. C. (1904). The mammalian cerebellum: Its lobes and fissures, Part 1. *J. Anat. Physiol.* **38,** 448–475.

Bolk, L. (1906). "Das Cerebellum der Saeugetiere." De Erven F. Bohn and Gustav Fischer, Jena.

Brodal, A. (1981). "Neurological Anatomy in Relation to Clinical Medicine." Third Edition. Oxford University Press.

Carpenter, M. B. (1976). "Human Neuroanatomy." The Williams & Wilkins Co., Baltimore, MD.

Collins, D. L., Neelin, P., Peters, T. M., and Evans, A. C. (1994). Automatic 3D intersubject registration of MR volumetric data in standardized Talairach space. *J. Comput. Assist. Tomogr.* **18,** 192–205.

Courchesne, E., Press, G. A., Murakami, J., Berthoty, D., Grafe, M., Wiley, C. A., and Hesselink, J. R. (1989), The cerebellum in sagittal plane—Anatomic-MR correlation: 1. The Vermis. *AJNR* **10,** 659–665.

Crosby, E. C., Humphrey, T., and Lauer, E. W. (1962). *In* "Correlative Anatomy of the Nervous System," pp. 188–192. Macmillan Press, New York.

Damasio, H., and Damasio, A. R. (1989). "Lesion Analysis in Neuropsychology." Oxford University Press, New York.

DeArmond, S. J., Fusco, M. M., and Devey, M. M. (1976). "A Photographic Atlas: Structure of the Human Brain." Oxford University Press, New York.

Dejerine, J. (1901). "Anatomie des Centres Nerveux, Tome 2." J. Rueff et Cie, Paris.

Desmond, J. E., Gabrieli, J. D., and Glover, G. H. (1998). Dissociation of frontal and cerebellar activity in a cognitive task: Evidence for a distinction between selection and search. *Neuroimage* **7,** 368–376.

Dow, R. S. (1942). The evolution and anatomy of the cerebellum. *Biol. Rev.* **17,** 179–220.

Doyon, J. (1997). Skill learning. *In* "The Cerebellum and Cognition. International Review of Neurobiology, Vol. 41" (J. D. Schmahmann, ed.), pp. 273–296. Academic Press, San Diego.

Edinger, L. (1909). Uber die Enteilung des Cerebellums. *Anatomischer Anzeiger* **35,** 319–323.

Fiez, J. A., and Raichle, M. E. (1997). Linguistic processing. *In* "The Cerebellum and Cognition. International Review of Neurobiology, Vol. 41" (J. D. Schmahmann, ed.), 233–254. Academic Press, San Diego.

Fix, J. D., and Punte, C. S. (1981). "Atlas of the Human Brain Stem and Spinal Cord." University Park Press, Baltimore.

Flatau, E., and Jacobsohn, L. (1899). "Handbuch der Anatomie und vergleichende Anatomie des Centralnervensystems der Saeugetiere. I. Makroskopischer Teil." S. Karger, Berlin.

Gray's Anatomy 38th Edition. (1995) pp.1033–1035. Churchill Livingston, New York.

Henle, J. (1901). "Grundriss der Anatomie des Menschen. Neu bearbeitet von F. Merkel. 4th Aufl. Atlas." Friedrich Vieweg & Sohn, Braunschweig.

His, W. (1895). Die anatomische Nomenklatur. Nomina anatomica. Separat-Abzug aus Archiv für Anatomie und Physiologie. Anatomische Abteilung. Supplement-Band. (Basle Nomina Anatomica of 1895; BNA.)

Holmes, C. J., Hoge, R., Collins, L., Woods, R., Toga, A. W., and Evans, A. C. (1998). Enhancement of MR images using registration for signal averaging. *J. Comput. Assist. Tomogr.* **22,** 324–333.

Ingvar, S. (1918). Zur Phylo- und Ontogenesae des Kleinhirns. *Folia Neuro-Biologica* **11,** 205–495.

Ingvar, S. (1928). Studies in neurology, I. The phylogenetic continuity of the central nervous system. *Bull. Johns Hopkins Hospital* **43,** 315–337.

International Anatomical Nomenclature Committee. (1956). "Nomina Anatomica." Revised by the I.A.N.C. appointed by the Fifth International Congress of Anatomists held at Oxford in 1950. Submitted to the Sixth International Congress of Anatomists, held in Paris, July 1955. Williams & Wilkins Co., Baltimore.

Jakob, A. (1928). Das Kleinhirn. *In* "Handbuch der mikroskopischen Antomie des Menschen, Vol. 4" (v. Möllendorff, ed.), pp. 674–916. Berlin: Springer.

Jansen, J., Brodal, A. (1958). Das Kleinhirn. *In* "Handbuch der mikroskopischen Antomie des Menschen, Vol. 4" (v. Möllendorff, ed.), pp. 1–323. Springer, Berlin.

Kretschmann, H. J., and Weinrich, W. (1992). "Cranial Neuroimaging and Clinical Neuroanatomy." Thieme, New York.

Kuithan, W. (1894). Die Entwicklung des Kleinhirns von Säugetieren, unter Ausschluß der Histogenese. *Sitzungsber. Ges. Morphol. Physiol.* **10,** 89–128.

Langelaan, J. W. (1919). On the development of the external form of the human cerebellum. *Brain* **42,** 130–170.

Larsell, O. (1934). Morphogenesis and evolution of the cerebellum. *Arch. Neurol. Psych.* **31,** 373–395.

Larsell, O. (1937). The cerebellum: A review and interpretation. *Arch. Neurol. Psych.* **38,** 580–607.

Larsell, O. (1947). The development of the cerebellum in man in relation to its comparative anatomy. *J. Comp. Neurol.* **87,** 85–129.

Larsell, O. (1953). The anterior lobe of the mammalian and the human cerebellum. *Anat. Rec.* **115,** 341.

Larsell, O. (1958). Lobules of the mammalian and human cerebellum. *Anat. Rec.* **130,** 329–330.

Larsell, O. (1967). "The Comparative Anatomy and Histology of the Cerebellum from Myxinoids through Birds" (J. Jansen, ed.). The University of Minnesota Press, Minneapolis.

Larsell, O. (1970). "The Comparative Anatomy and Histology of the Cerebellum from Monotremes through Apes" (J. Jansen, ed.). The University of Minnesota Press, Minneapolis.

Larsell, O., and Jansen, J. (1972). "The Comparative Anatomy and Histology of the Cerebellum. The Human Cerebellum, Cerebellar Connections, and Cerebellar Cortex." The University of Minnesota Press, Minneapolis.

Lorenson, W. E., Cline, H. E. (1987). Marching cubes: A high resolution surface reconstruction algorithm. *Comp. Graphics* **4,** 163–169.

Loyning, Y., and Jansen, J. (1955). A note on the morphology of the human cerebellum. *Acta Anat.* **25,** 309–318.

Madigan, J. C., and Carpenter, M. B. (1971). "Cerebellum of the Rhesus Monkey: Atlas of Lobules, Laminae, and Folia, in Sections." University Park Press, Baltimore.

Malacarne, M. V. G. (1776). "Nuova esposizione della vera struttura del cervelletto umano." G M Briolo, Torino.

Meckel, J. F. (1838). "Manual of Descriptive and Pathological Anatomy. Volume II." Translated from German into French by A. J. L. Jourdan and G. Breschet. Translated from French into English by A. S. Doane. E. Henderson, Old Bailey, London.

Miller, R. A., and Burack, E. (1977). "Atlas of the Central Nervous System in Man." Second edition. Williams and Wilkins, Baltimore.

Parsons, L. M., and Fox, P. T. (1997). Sensory and cognitive functions. *In* "The Cerebellum and Cognition. International Review of Neurobiology, Vol. 41" (J. D. Schmahmann, ed.), pp. 255–272. Academic Press, San Diego.

Paxinos, G., Huang, X.-F., and Toga, A. W. (2000). "The Rhesus Monkey Brain in Stereotaxic Coordinates." Academic Press, San Diego.

Press, G. A., Murakami, J., Courchesne, E., Berthoty, D., Grafe, M., Wiley, C. A., and Hesselink, J. R. (1989). The cerebellum in sagittal plane—Anatomic-MR correlation. 2. The cerebellar hemispheres. *AJNR* **10,** 667–676.

Press, G. A., Murakami, J., Courchesne, E., Grafe, M., and Hesselink, J. R. (1990). The cerebellum. 3. Anatomic-MR correlation in the coronal plane. *AJNR* **11,** 41–50.

Riley, H. A. (1929). The mammalian cerebellum. A comparative study of the arbor vitae and folial pattern. *Res. Publ. Ass. Res. Nervous Mental Dis.* **6,** 37–192.

Riley, H. A. (1930). The lobules of the mammalian cerebellum and cerebellar nomenclature. *Arch. Neurol. Psych.* **24,** 227–256.

Riley, H. A. (1960). "An Atlas of the Basal Ganglia, Brain Stem and Spinal Cord Based on Myelin-stained Material." Hafner, New York.

Roberts, M. P., Hanaway, J., and Morest, D. K. (1987). "Atlas of the Human Brain in Sections." Lea & Febiger, Philadelphia.

Schäfer, E. A., and Symington, J. (1908). Neurology. *In* "Quain's Elements of Anatomy, 11th Edition," pp. 167–201. Longmans, Green & Co, London.

Schmahmann, J. D. (ed.) (1997). "The Cerebellum and Cognition: International Review of Neurobiology, Vol. 41." Academic Press, San Diego.

Schwalbe, G. A. (1881). "Lehrbuch der Neurologie." Eduard Besold, Erlangen.

Smith, G. E. (1902). The primary subdivision of the mammalian cerebellum. *J. Anat. Physiol.* **36,** 381–385.

Stroud, B. B. (1895). The mammalian cerebellum. *J. Comp. Neurol.* **5,** 71–118.

Talairach, J., and Tournoux, P. (1988). "Co-Planar Stereotaxic Atlas of the Human Brain. 3-Dimensional Proportional System: An Approach to Cerebral Imaging" (Mark Rayport, trans.). Thieme Medical Publishers, Inc., New York.

Toga, A. W., Ambach, K., Quinn, B., Hutchin, M., and Burton, J. S. (1994). Postmortem anatomy from cryosectioned whole human brain. *J. Neurosci. Methods* **54,** 239–252.

Toga, A. W., Goldkorn, A., Ambach, K., Chao, K., Quinn, B. C., and Yao, P. (1997). Postmortem cryosectioning as an anatomic reference for human brain mapping. *Comput. Med. Imaging Graph.* **21,** 131–141.

Vicq-d'Azyr, F. (1786). "Traité d'anatomie et de physiologie, avec des planches coloriées: Tome premier." Francois Ambrose Didot l'aîné, Paris.

Voogd, J., Feirabend, H. K. P., Schoen, H. J. R. (1990). Cerebellum and precerebellar nuclei. *In* "The Human Nervous System." (G. Paxinos, ed.), pp.321–386. Academic Press, New York.

Waddington, M. M. (1984). "Atlas of Human Intracranial Anatomy," Academic Books, Rutland, VT.

Ziehen, T. (1934). Centralnervensystem. *In* "Handbuch der Anatomie," pp. 1230–1289. Verlag von Gustav Fischer, Jena.

IMAGES

Sagittal Series

X = 54

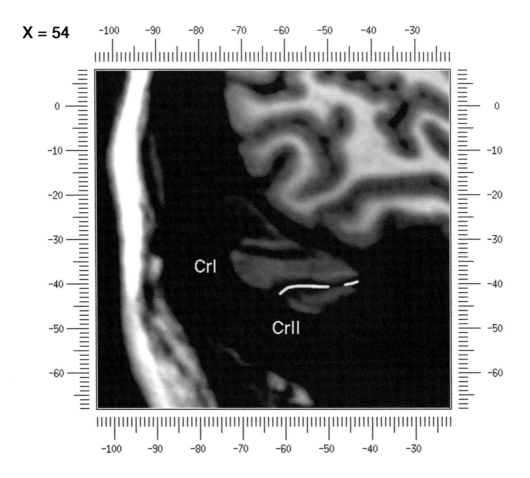

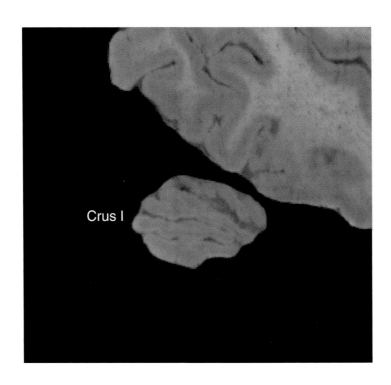

X = 52

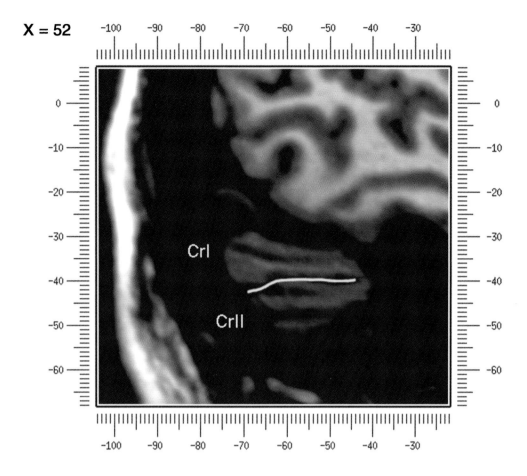

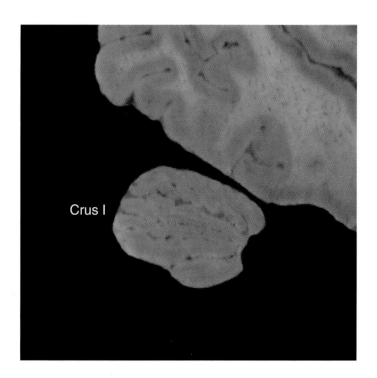

X = 50

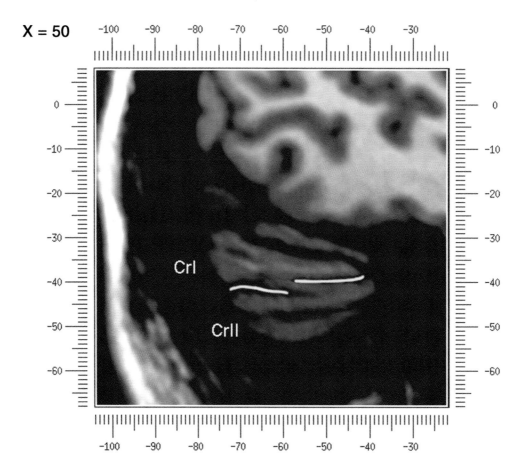

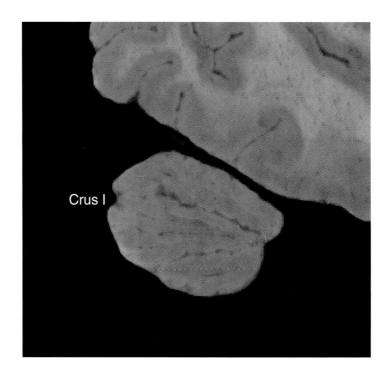

X = 48

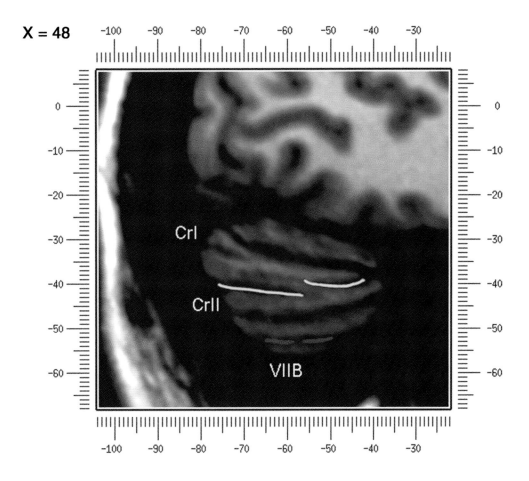

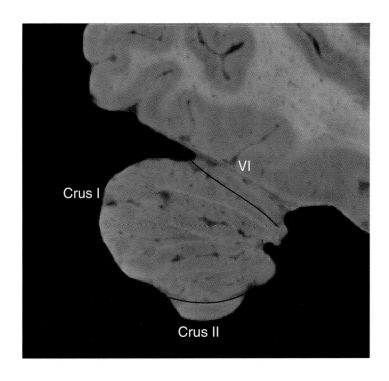

X = 46

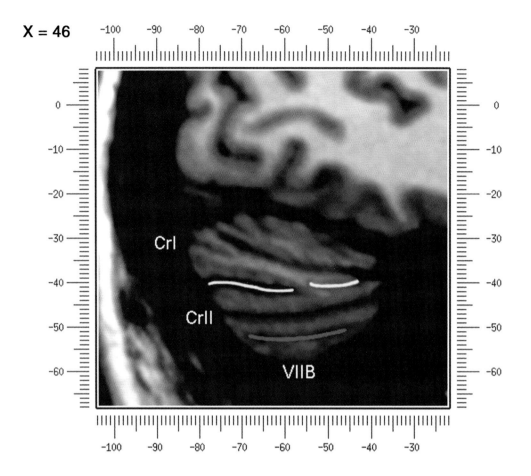

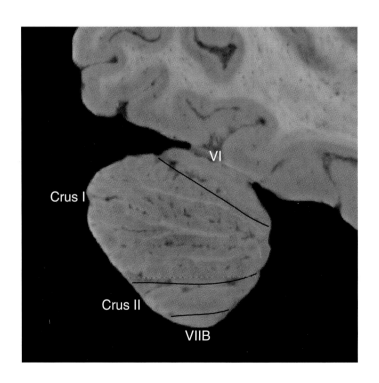

X = 44

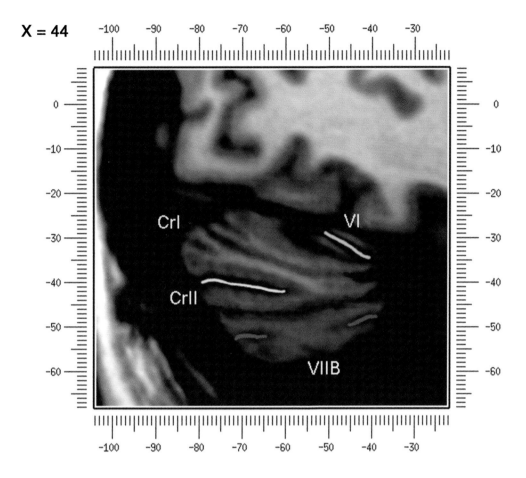

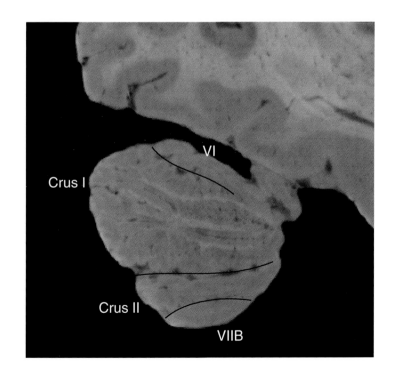

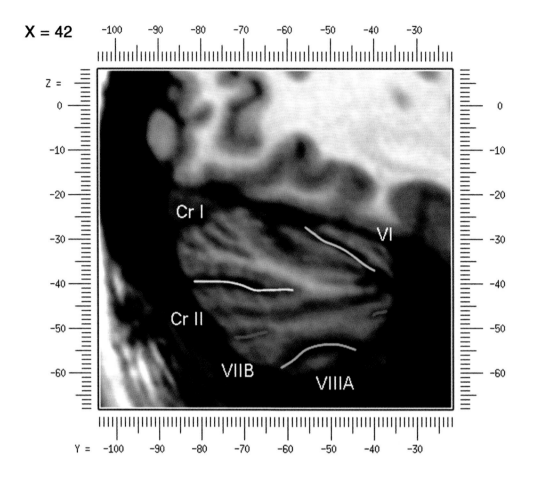

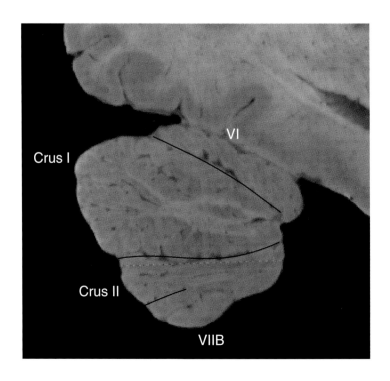

X = 40

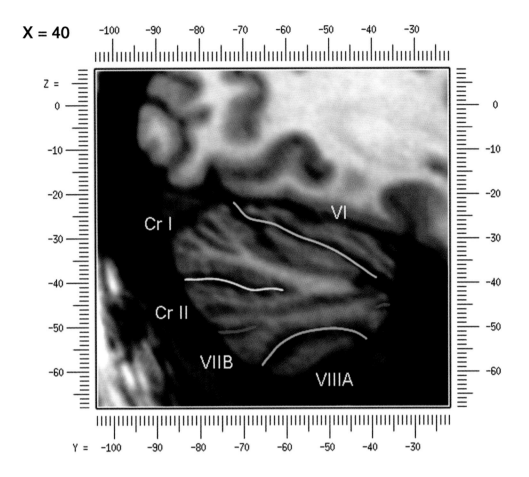

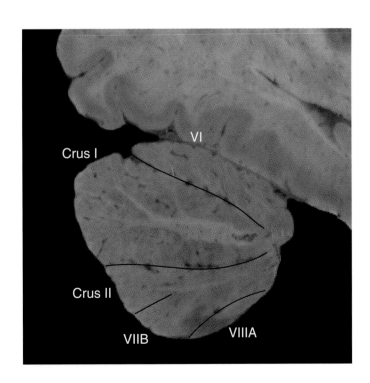

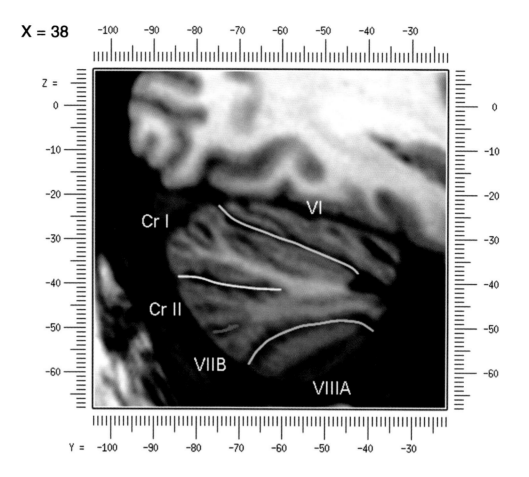

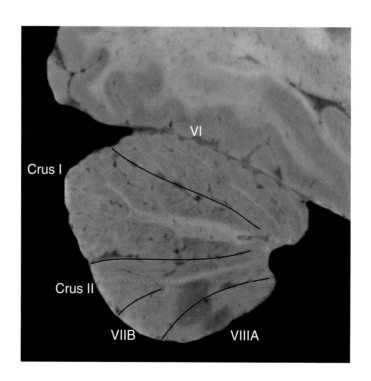

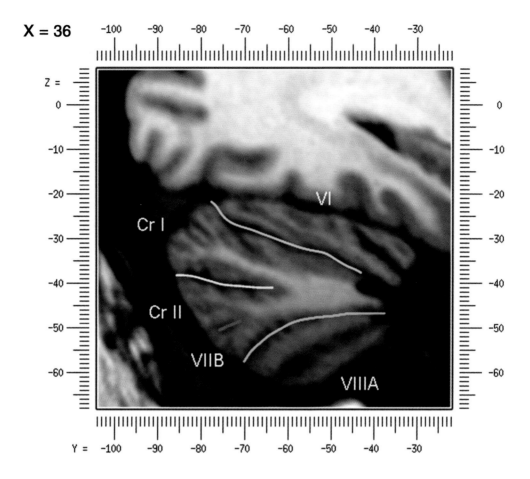

X = 36

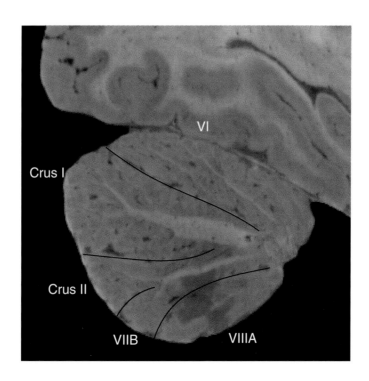

X = 34

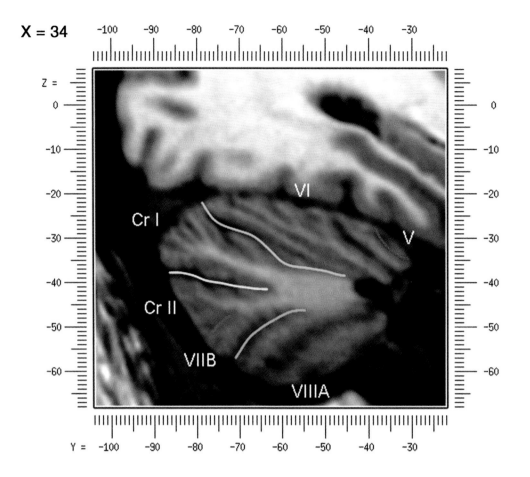

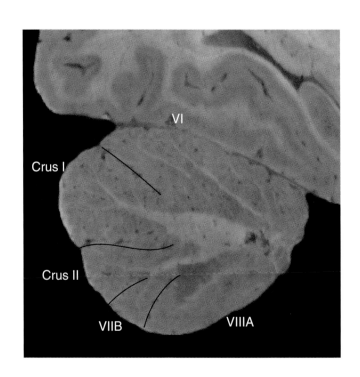

X = 32

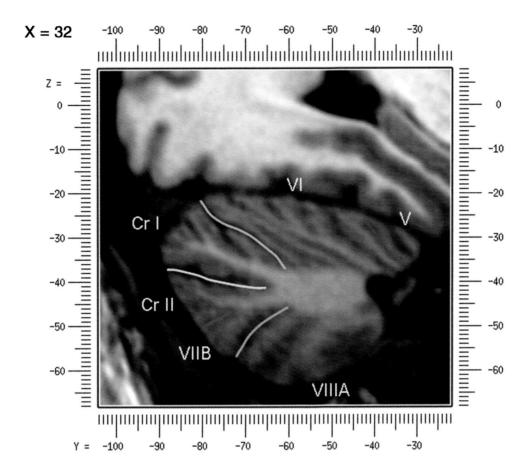

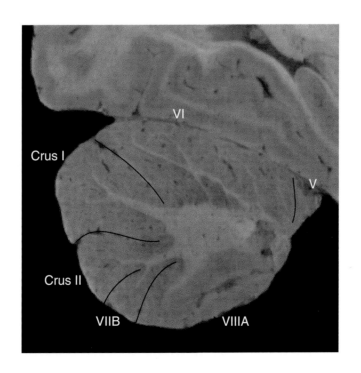

X = 30

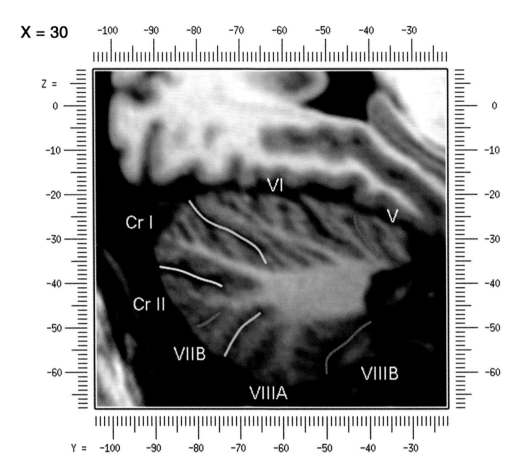

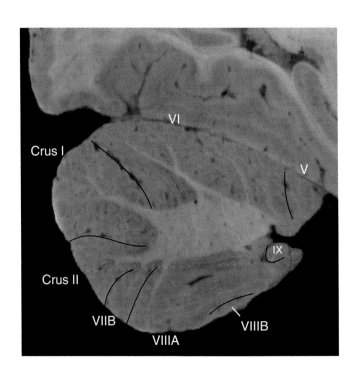

X = 28

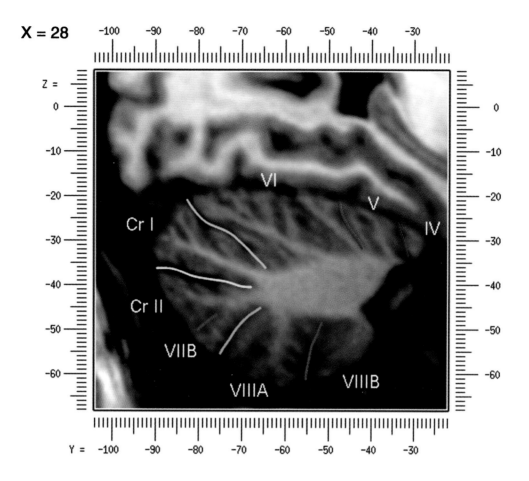

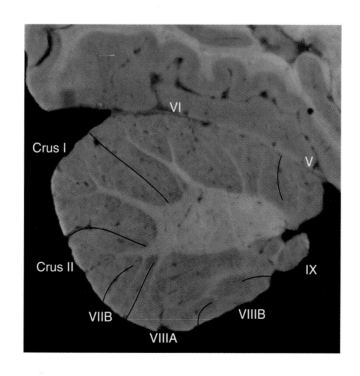

X = 26

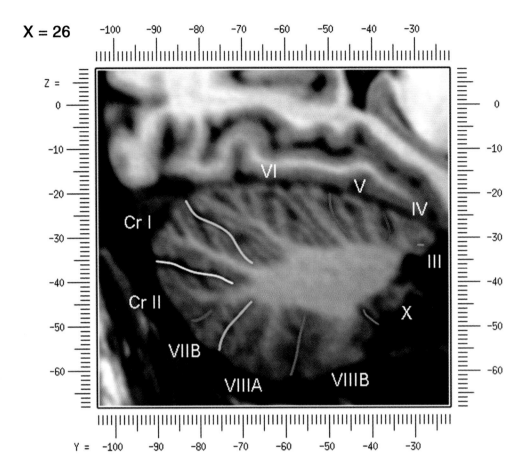

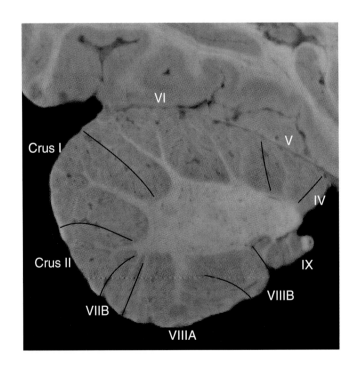

X = 24

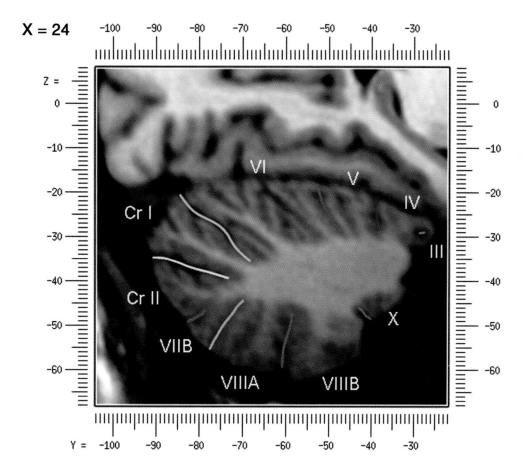

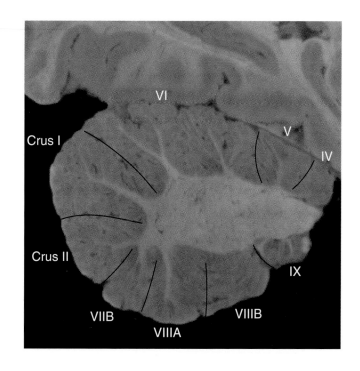

X = 22

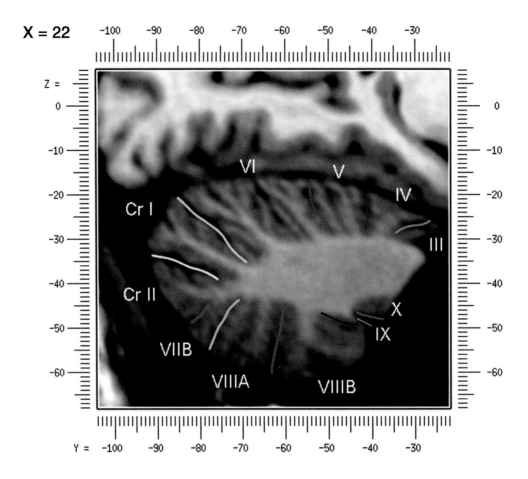

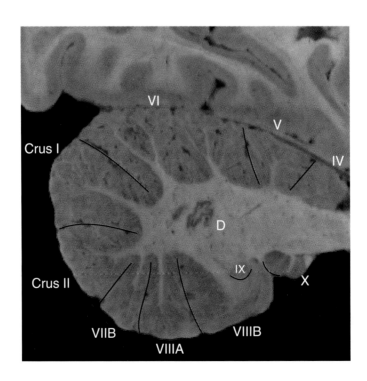

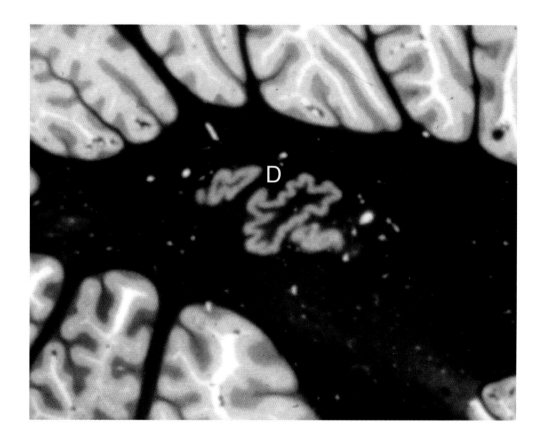

Myelin stain.
D= Dentate Nucleus

X = 20

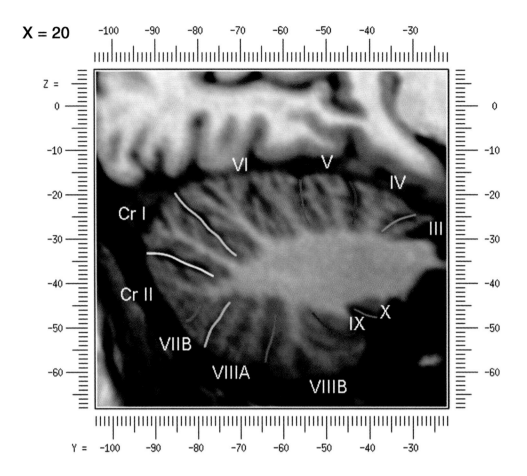

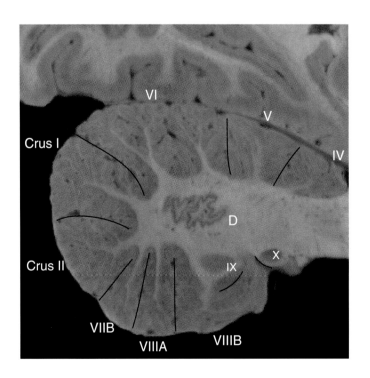

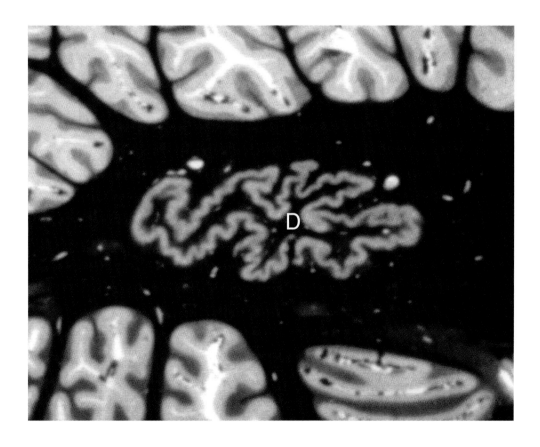

Myelin stain.
D= Dentate Nucleus

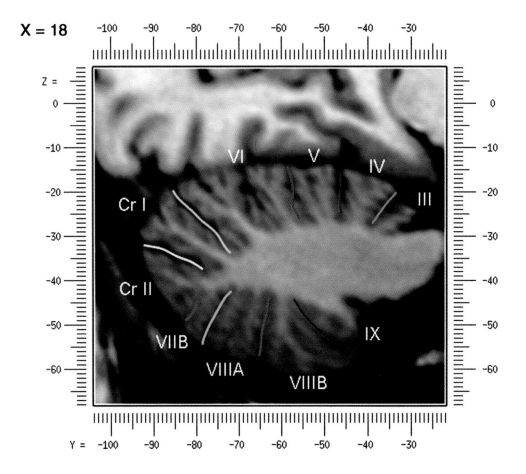

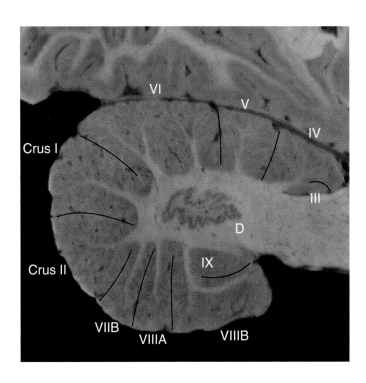

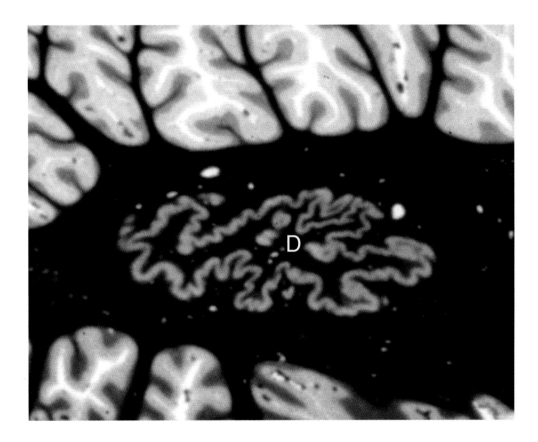

Myelin stain.
D= Dentate Nucleus

X = 16

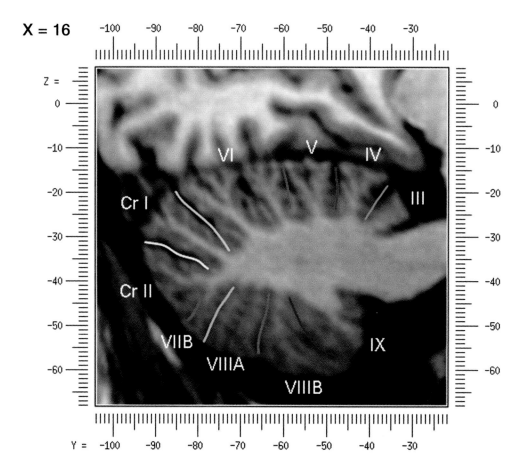

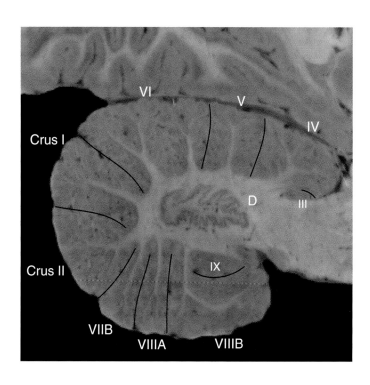

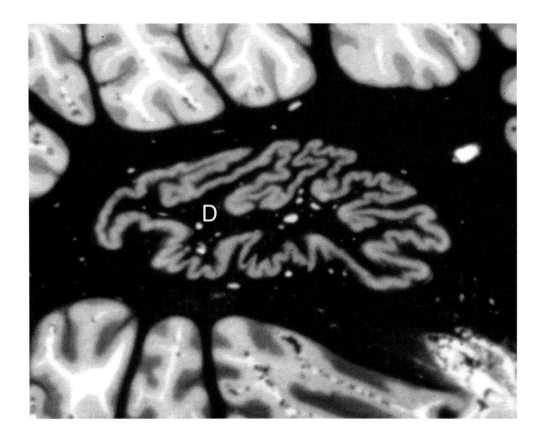

Myelin stain.
D= Dentate Nucleus

X = 14

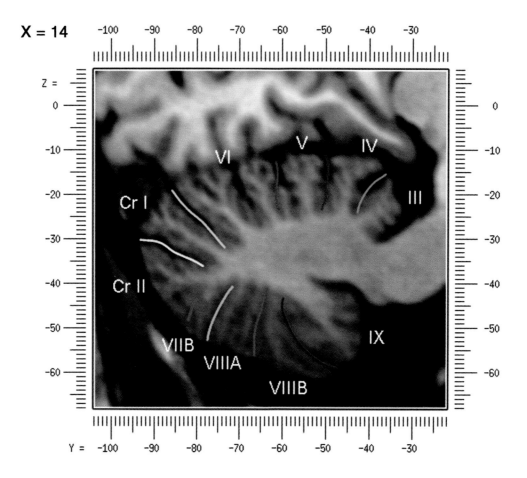

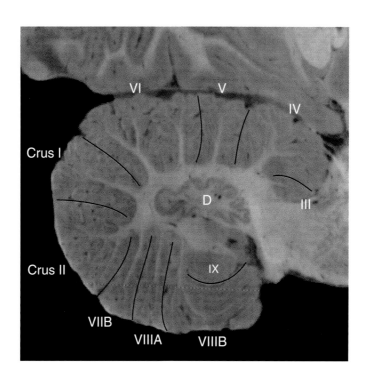

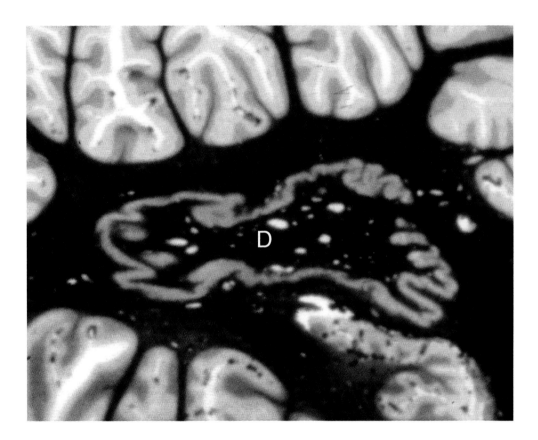

Myelin stain.
D= Dentate Nucleus

X =12

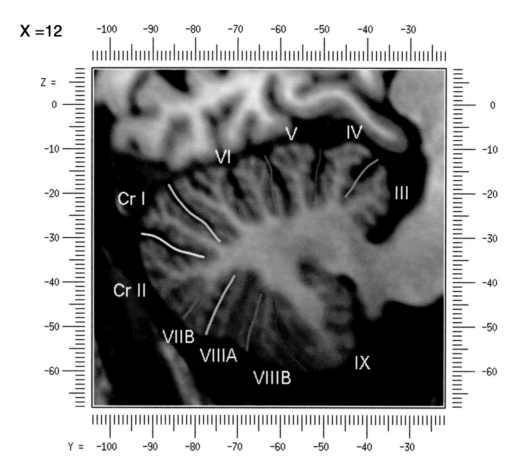

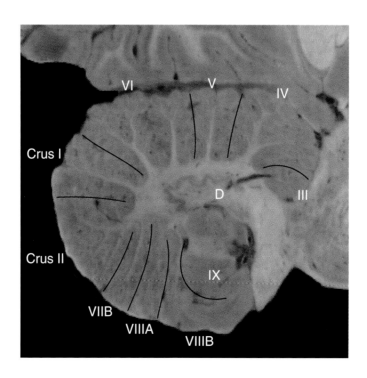

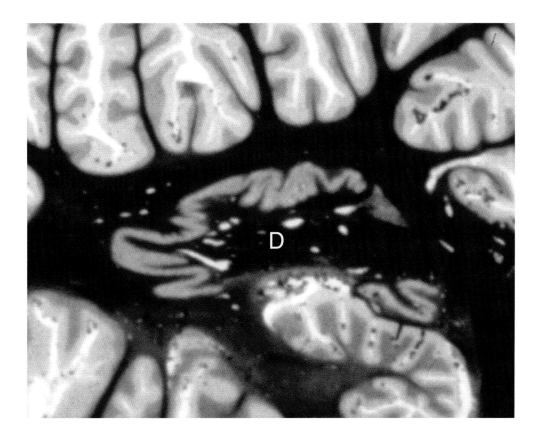

Myelin stain.
D= Dentate Nucleus

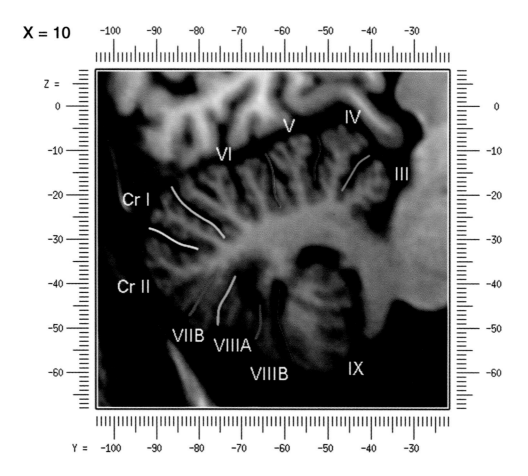

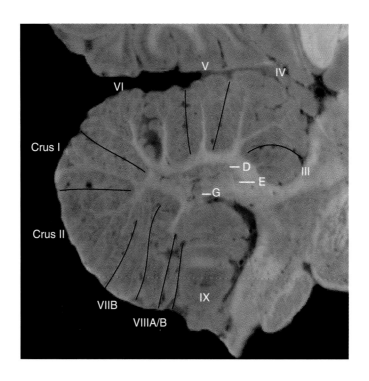

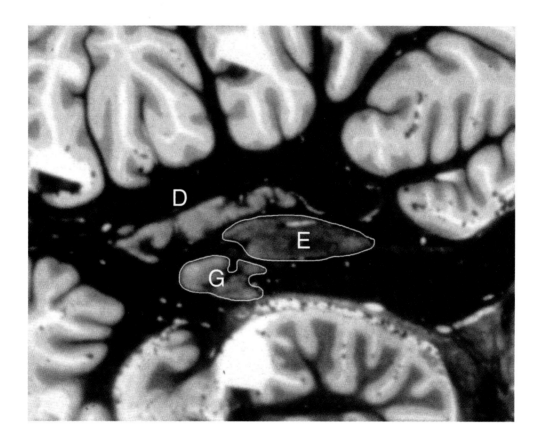

Myelin stain.
D= Dentate Nucleus, E= Emboliform Nucleus, G= Globose Nucleus

X = 8

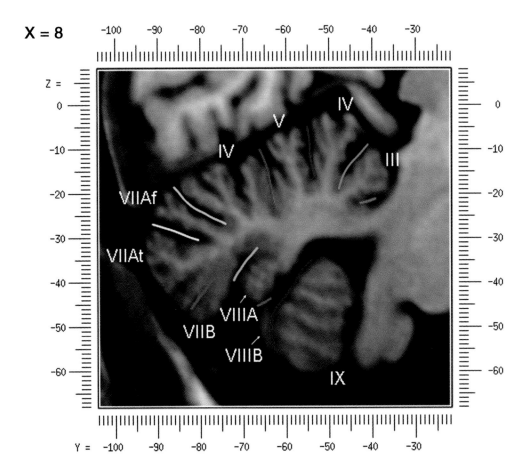

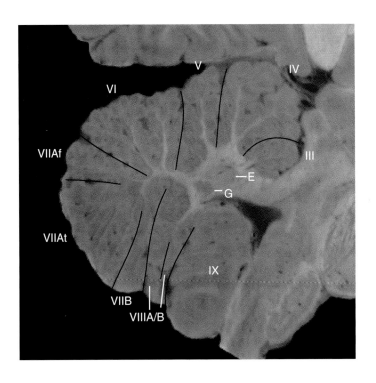

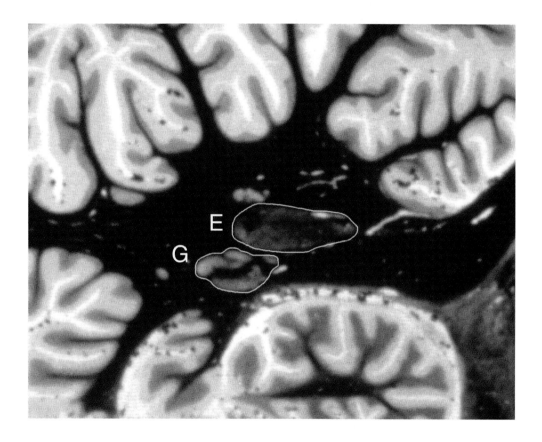

Myelin stain.
E= Emboliform Nucleus, G= Globose Nucleus

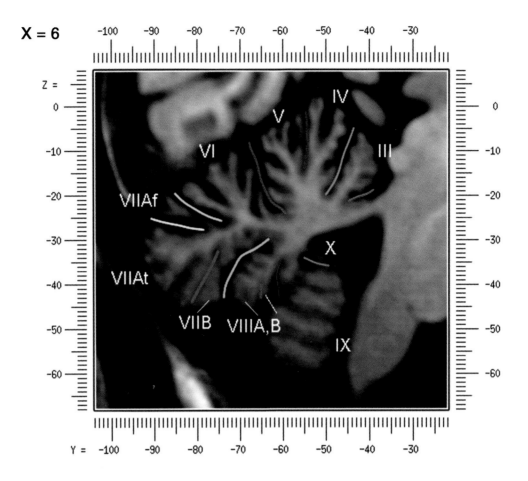

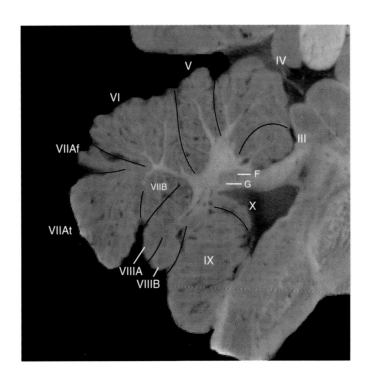

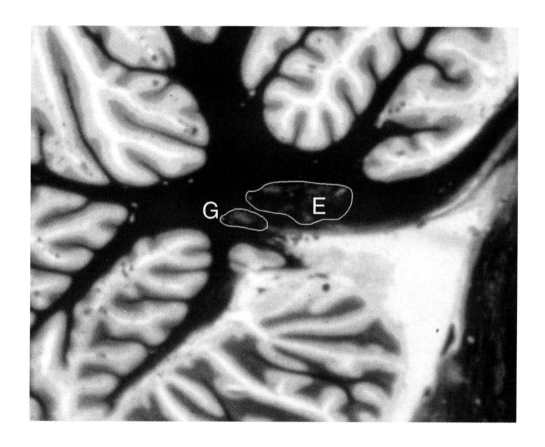

Myelin stain.
E= Emboliform Nucleus, G= Globose Nucleus

X = 4

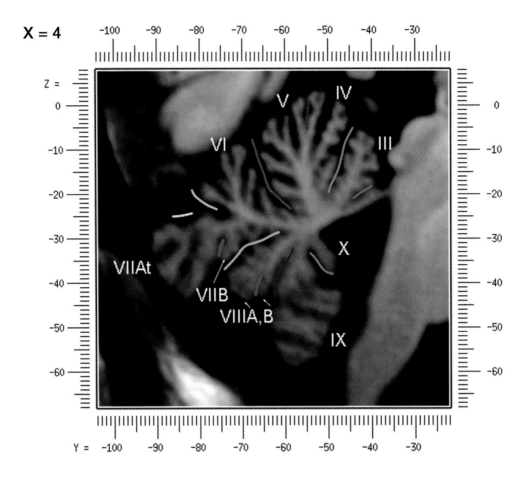

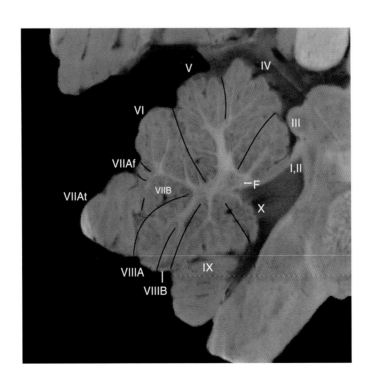

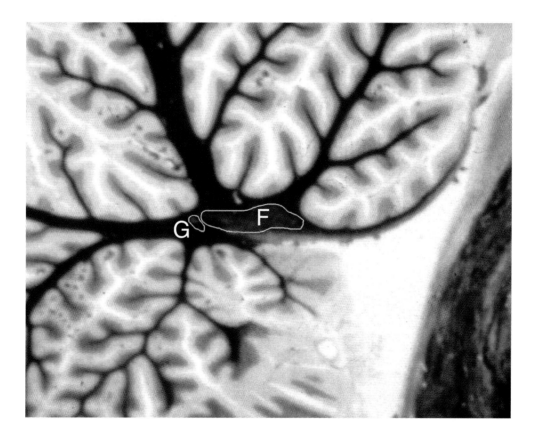

Myelin stain.
F=Fastigial Nucleus, G= Globose Nucleus

X = 2

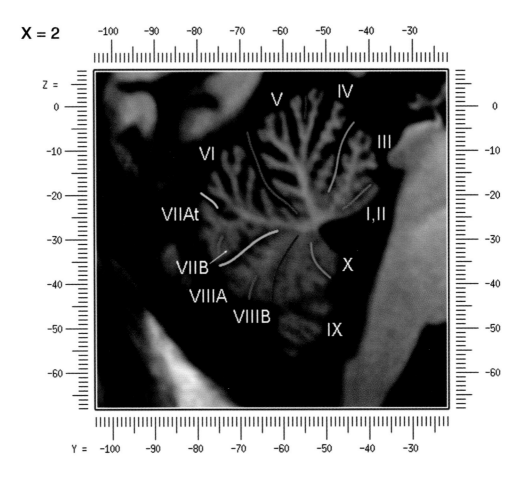

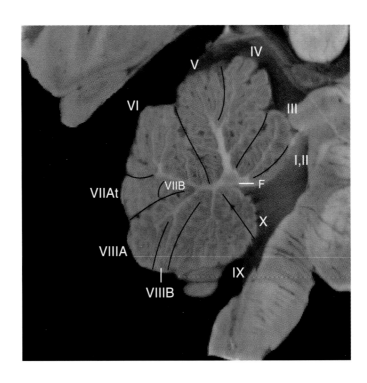

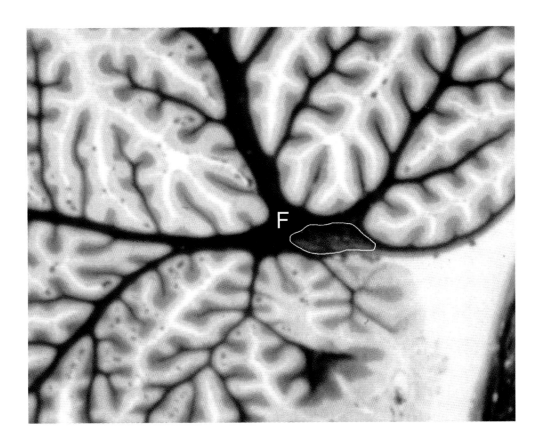

Myelin stain.
F=Fastigial Nucleus

X = 0

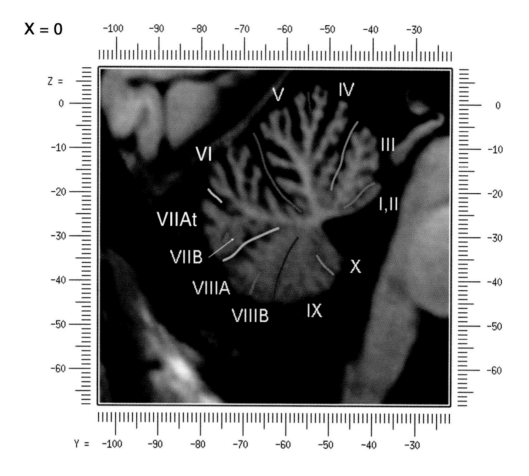

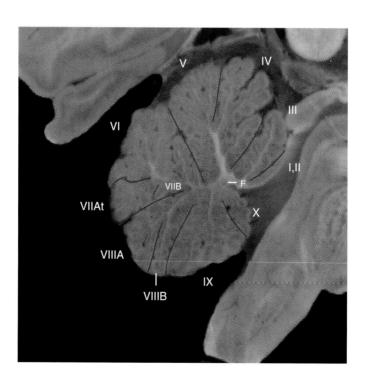

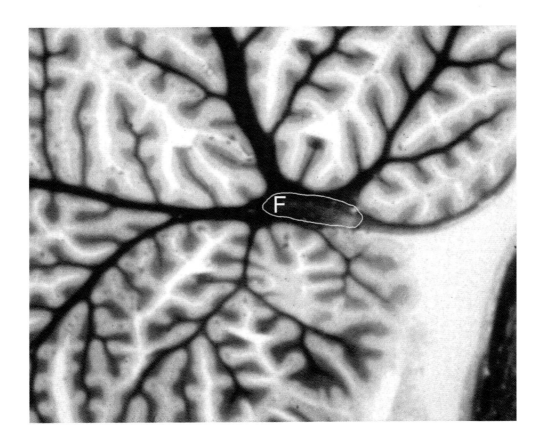

Myelin stain.
F=Fastigial Nucleus

X = -2

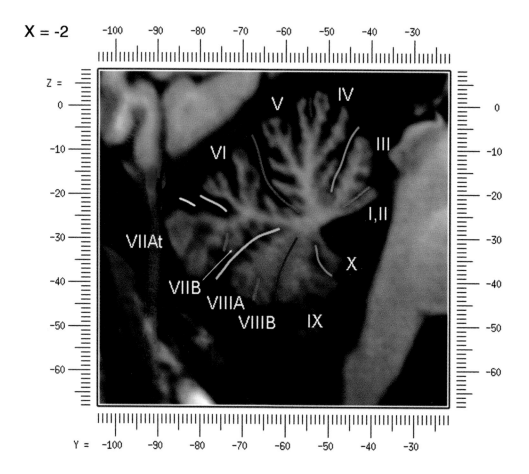

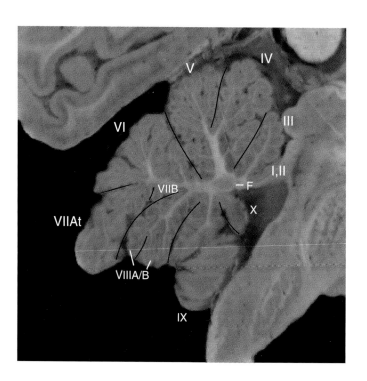

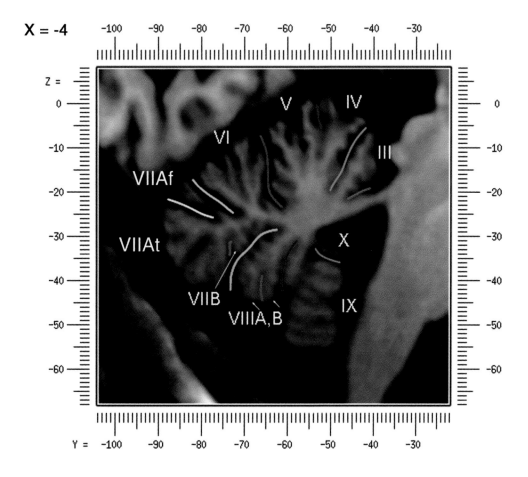

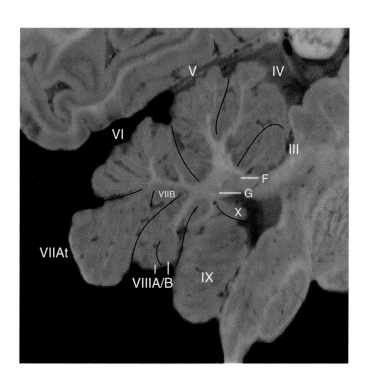

X = -6

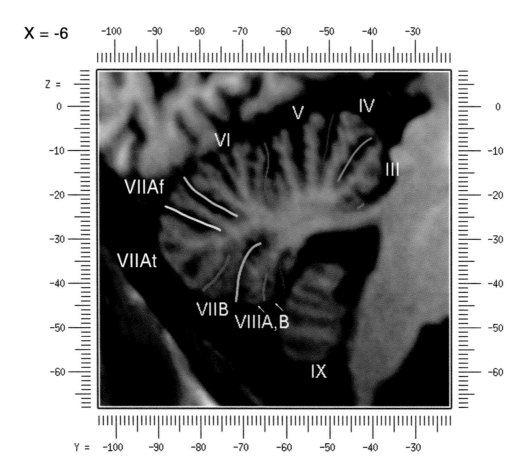

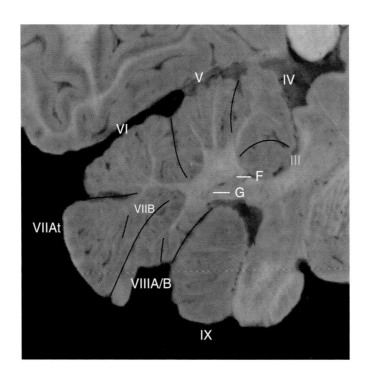

X = -8

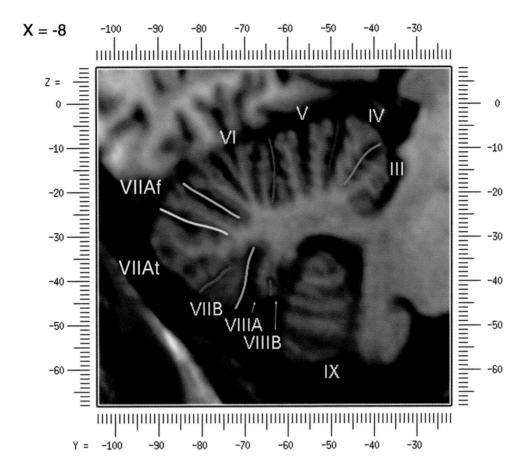

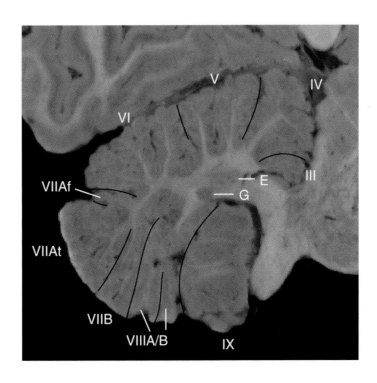

X = -10

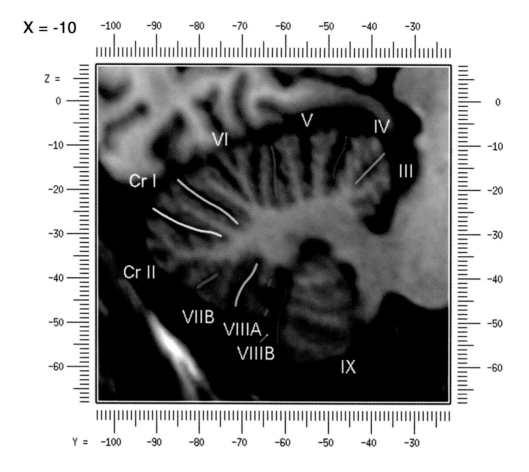

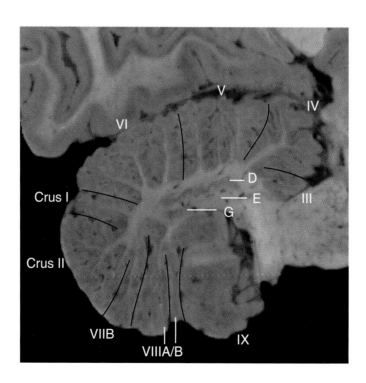

X = -12

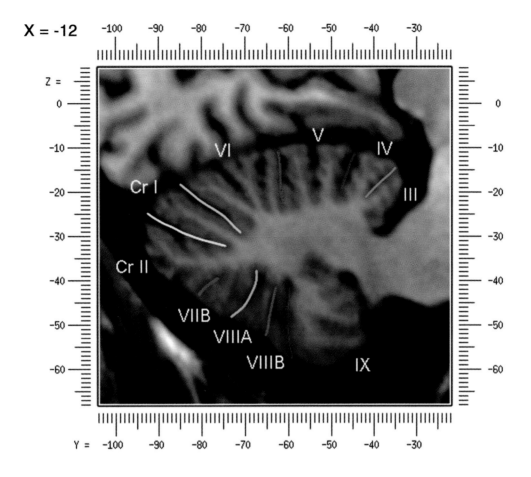

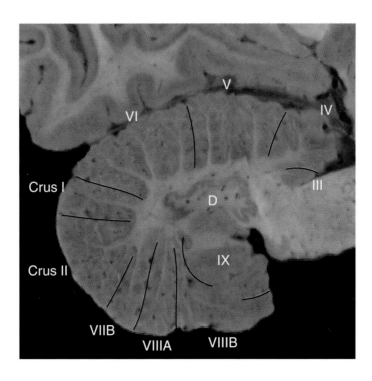

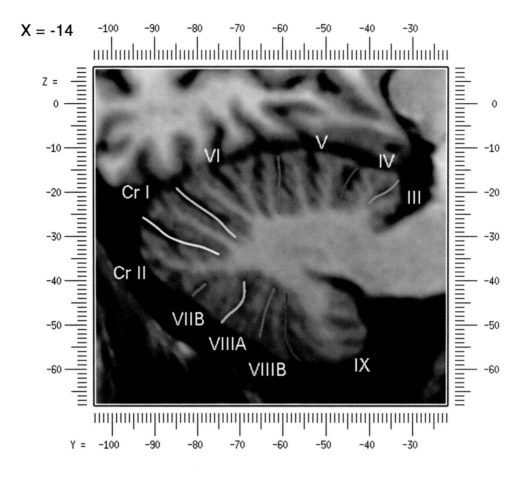

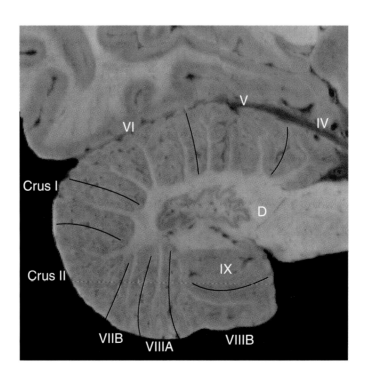

X = -16

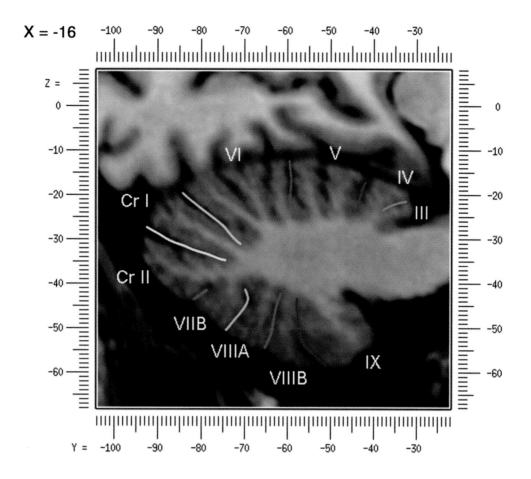

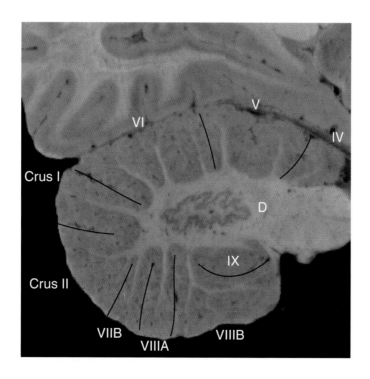

X = -18

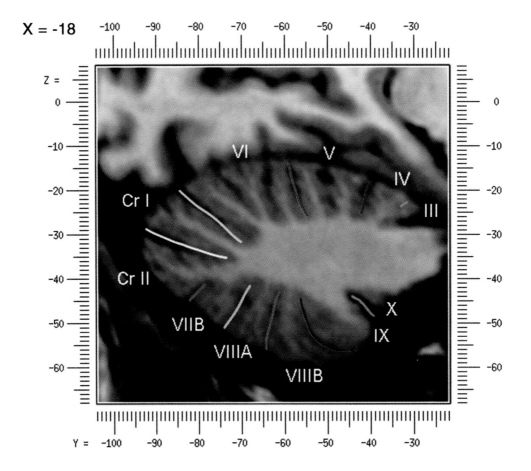

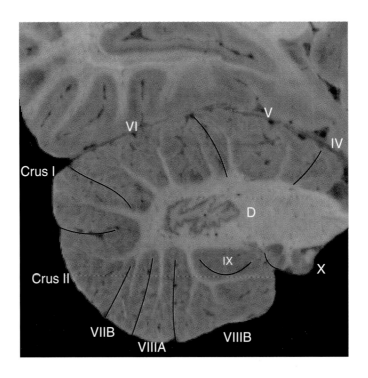

X = -20

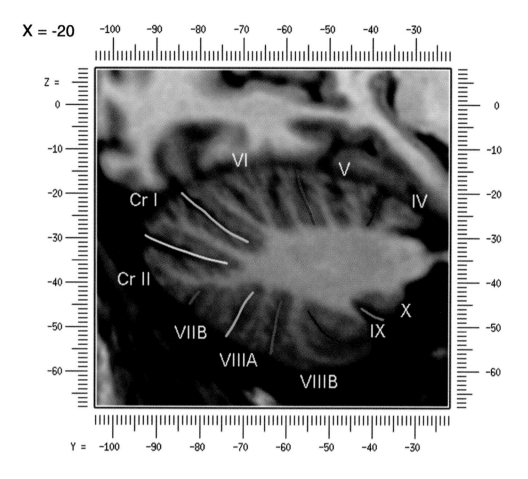

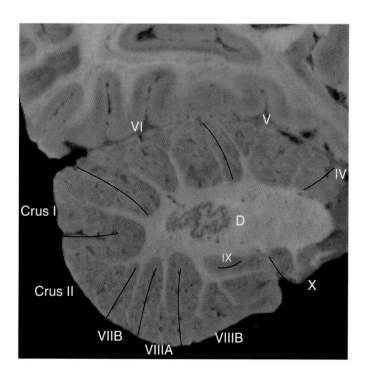

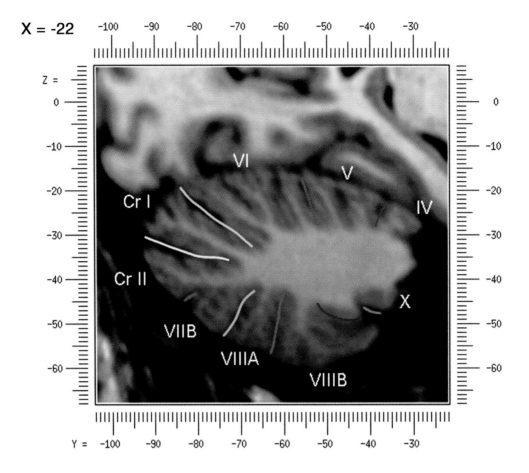

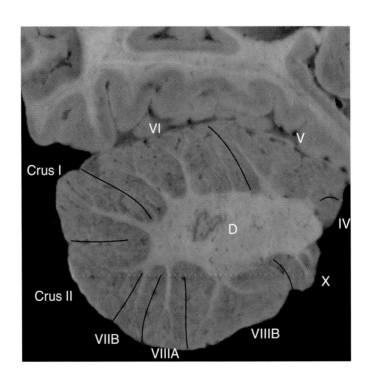

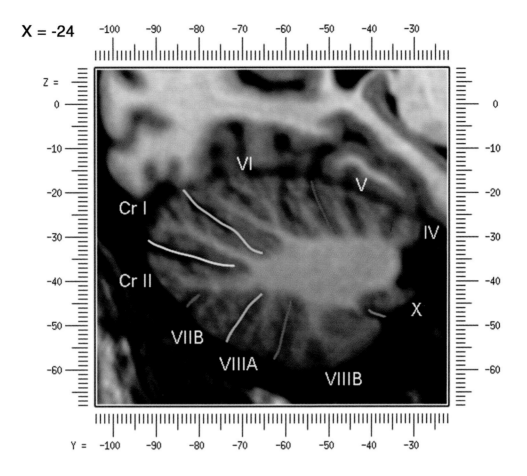

X = -24

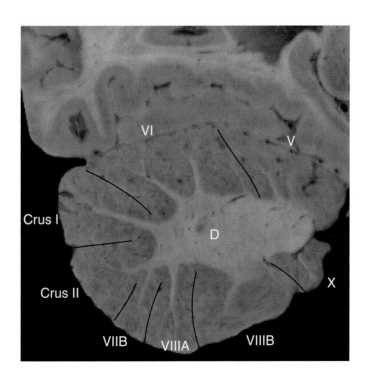

X = -26

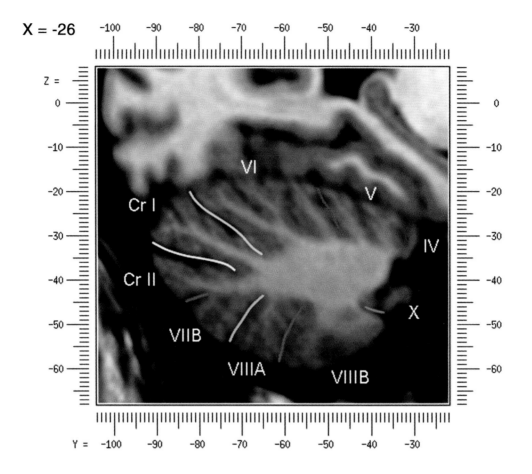

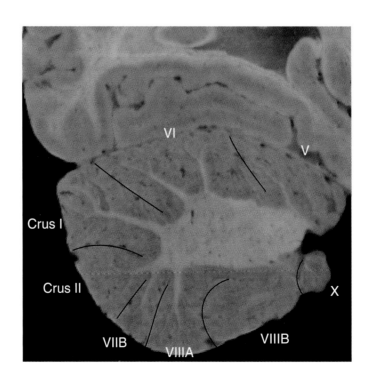

X = -28

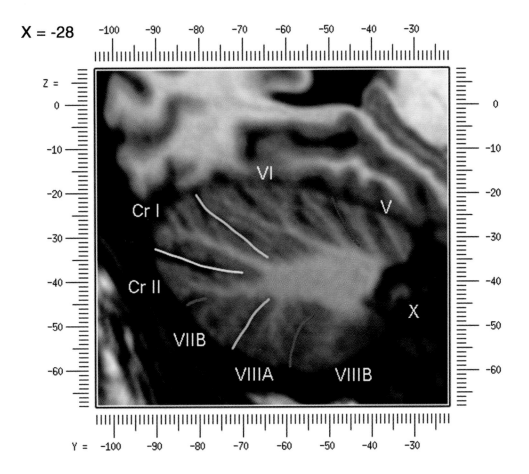

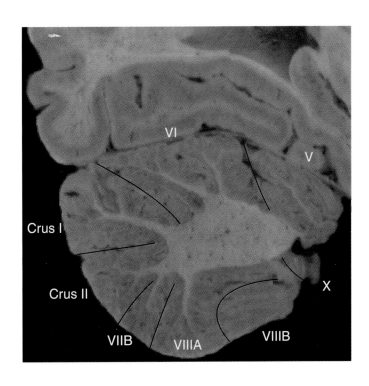

X = -30

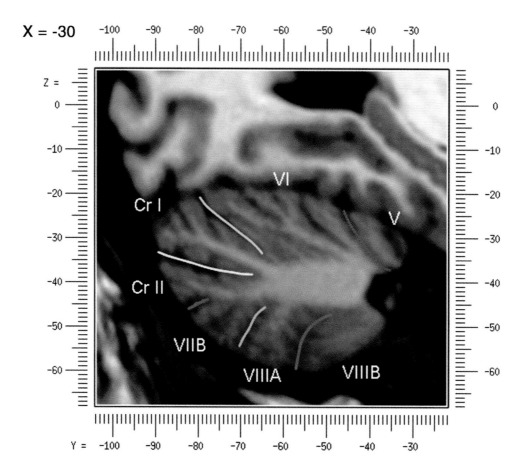

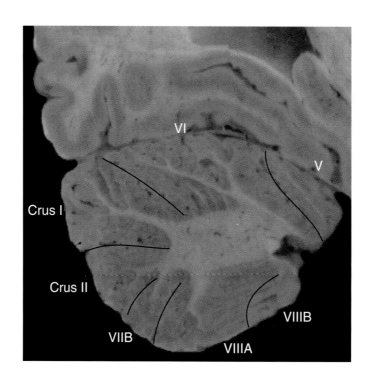

X = -32

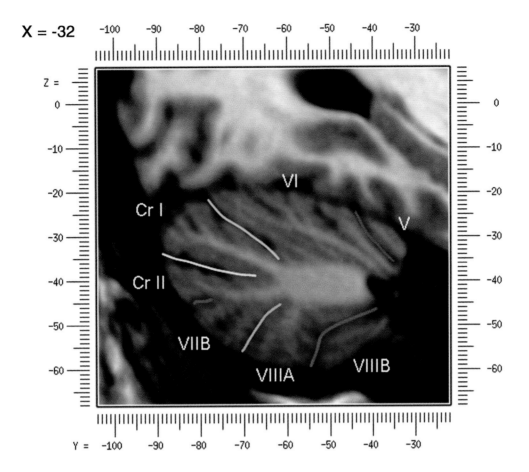

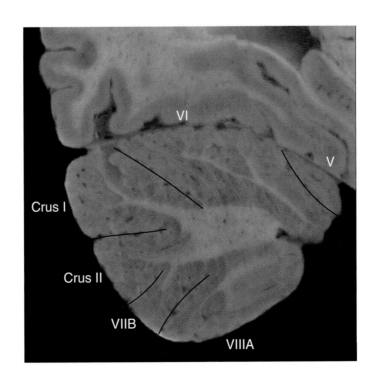

X = -34

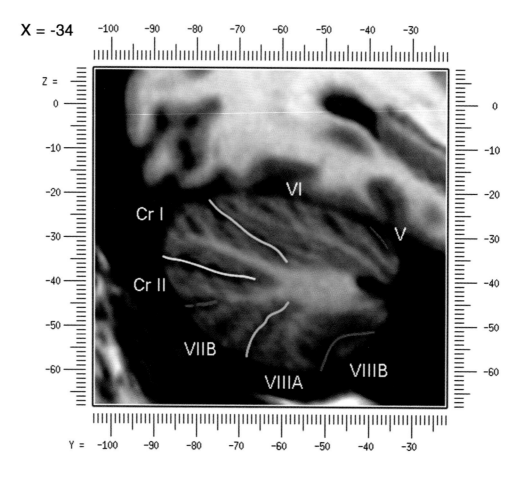

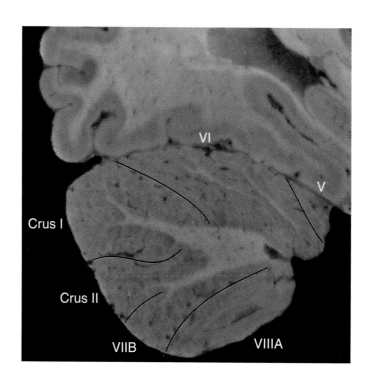

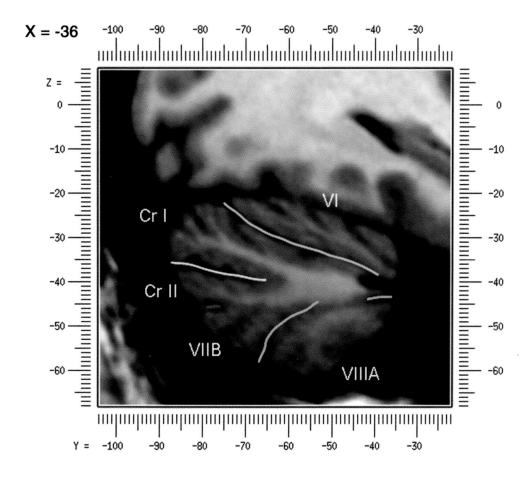

X = -36

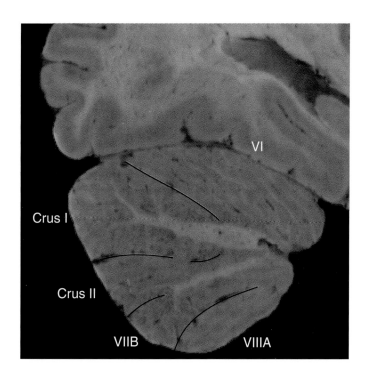

X = -38

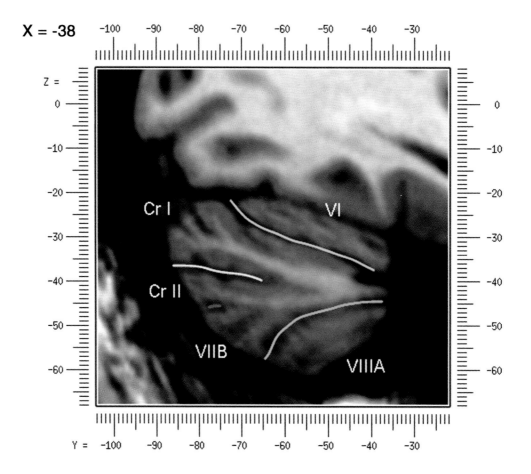

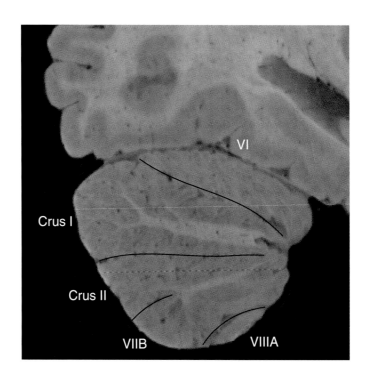

X = -40

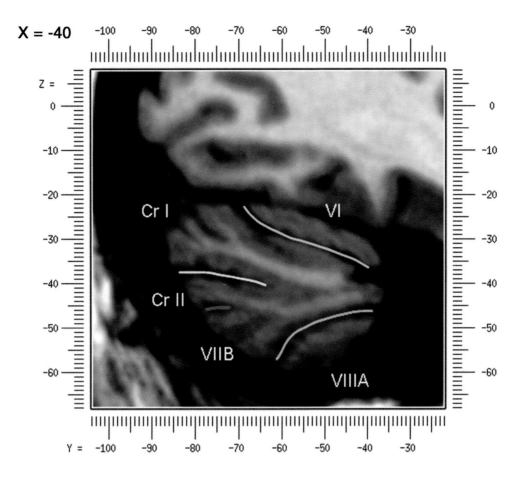

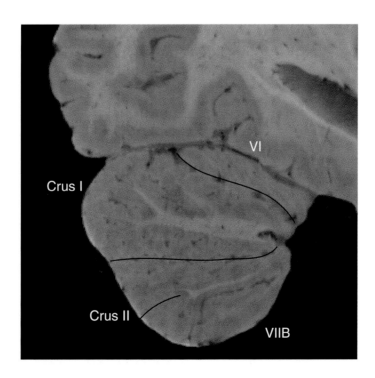

X = -42

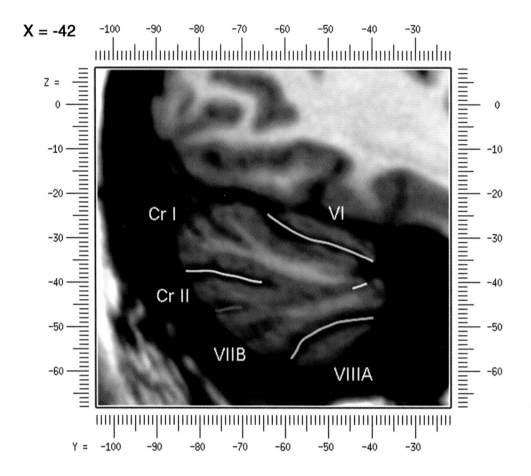

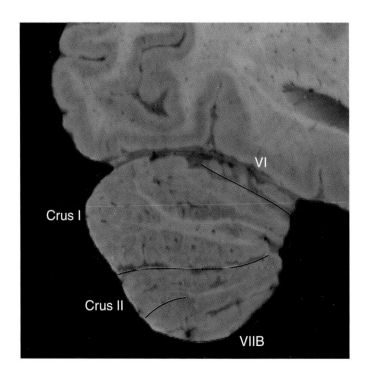

X = -44

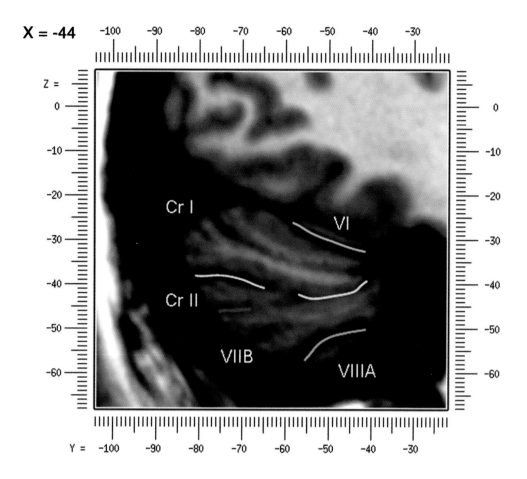

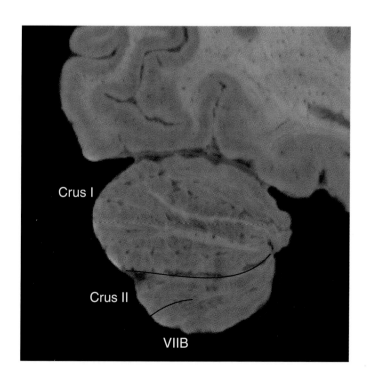

X = -46

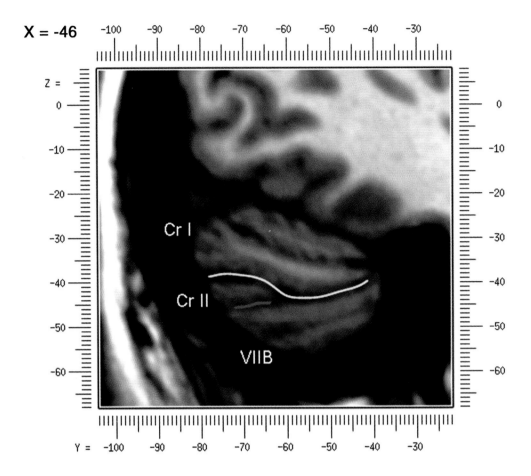

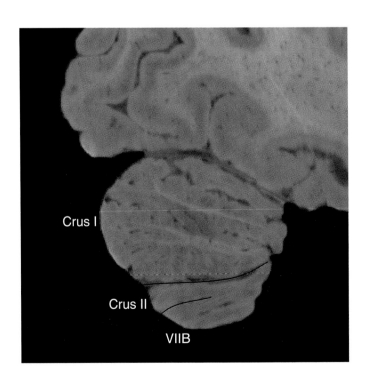

X = -48

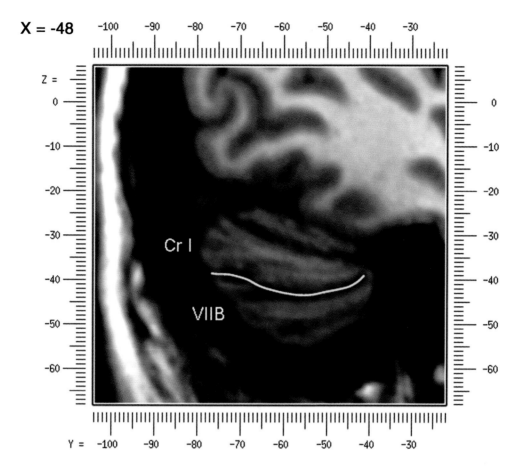

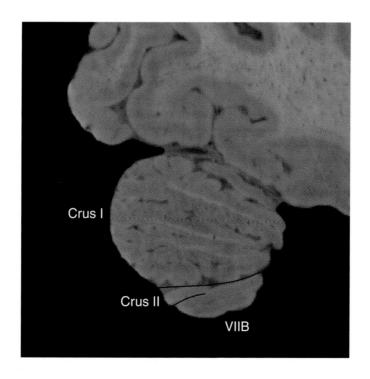

X = -50

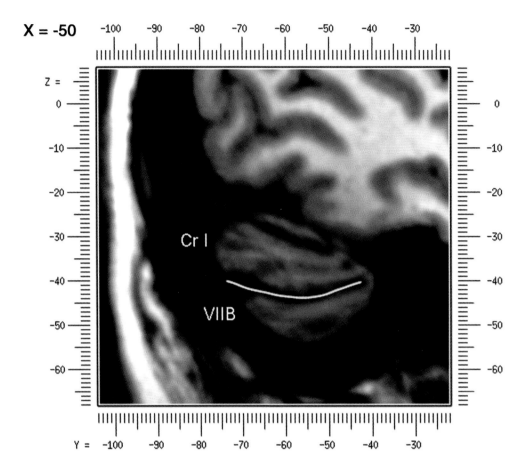

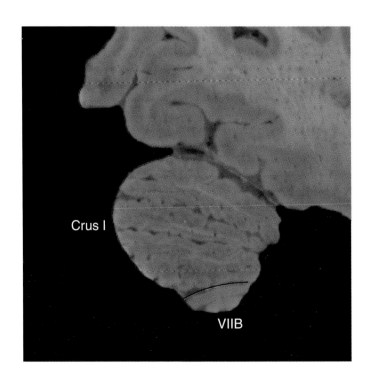

X = -52

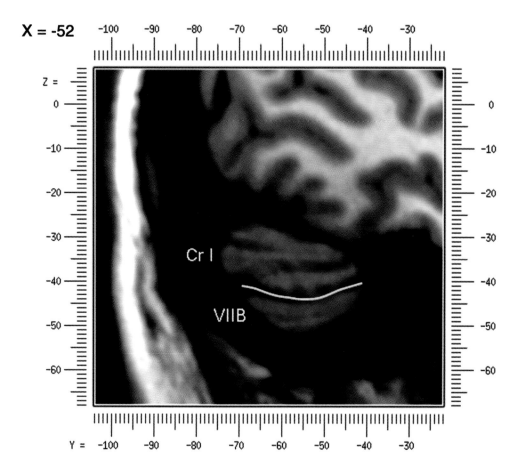

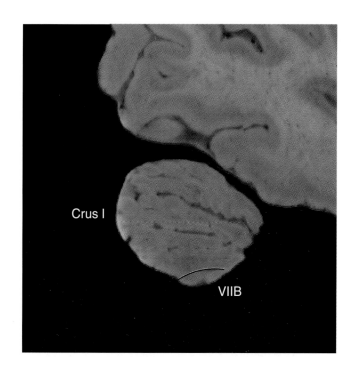

X = -54

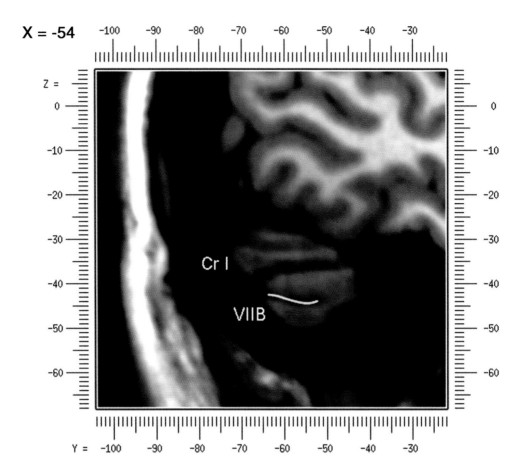

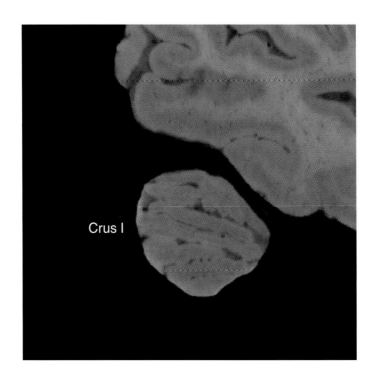

X = -56

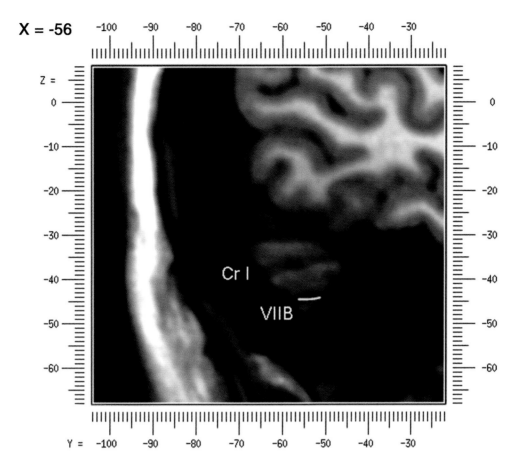

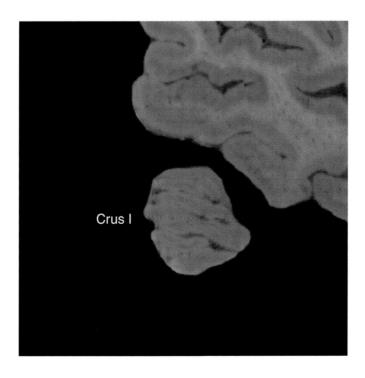

Coronal Series

Y = -32

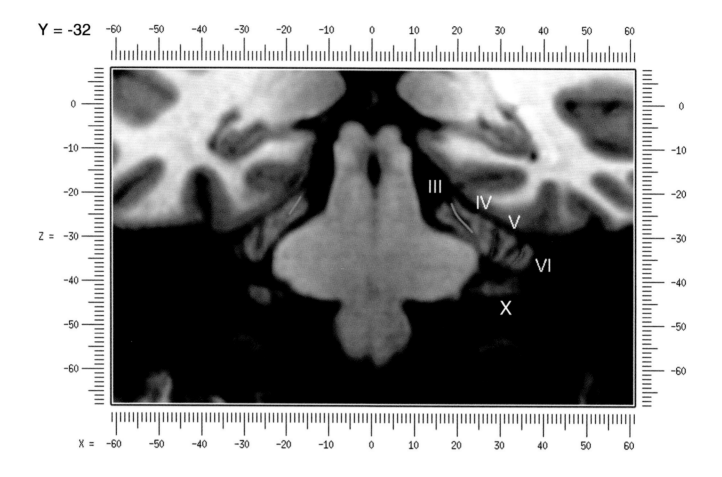

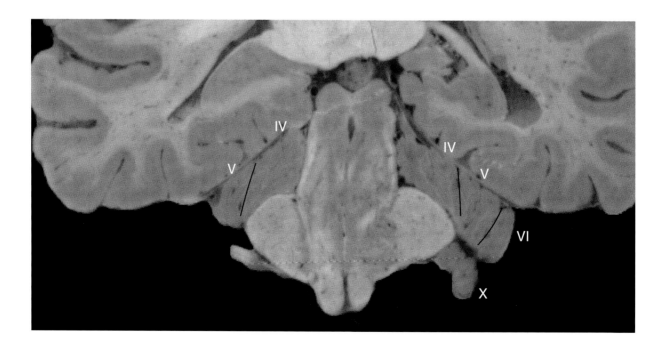

Y = -34

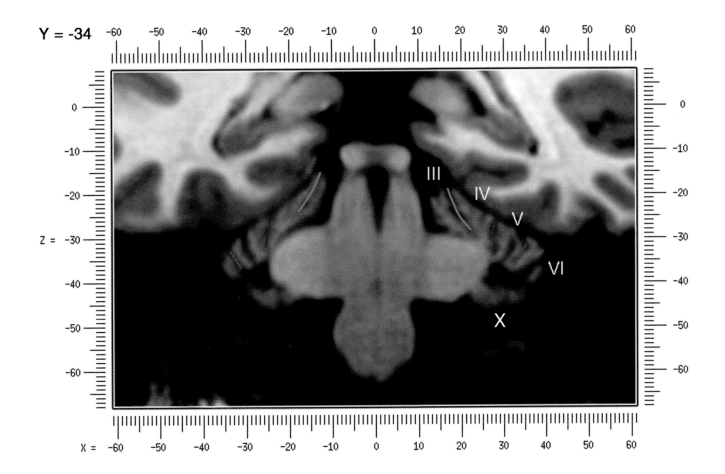

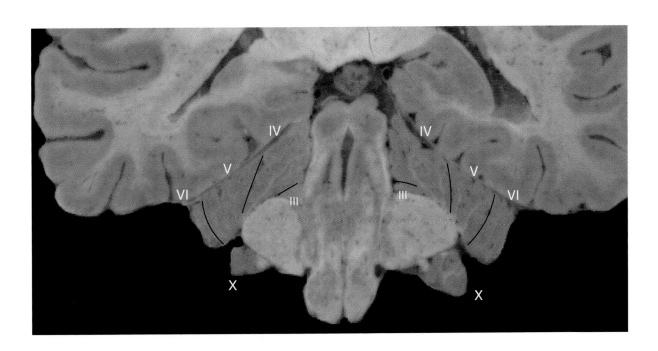

Y = -36

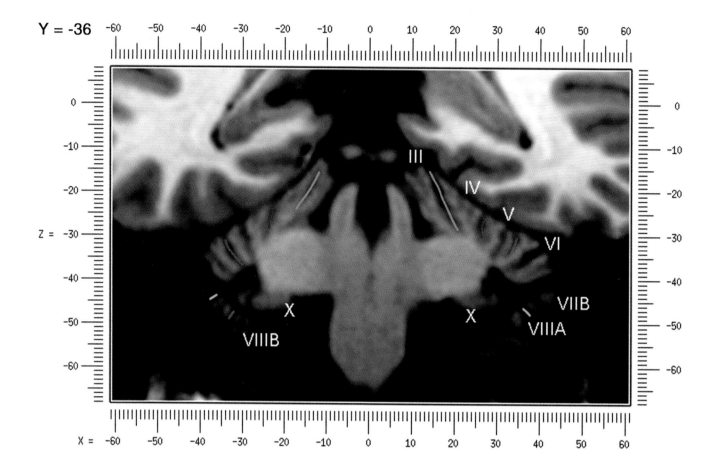

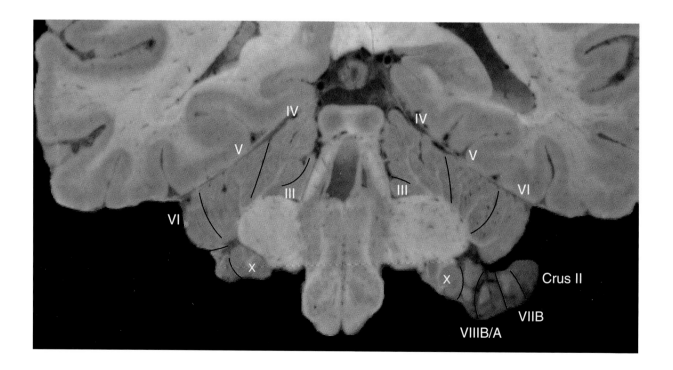

Y = -38

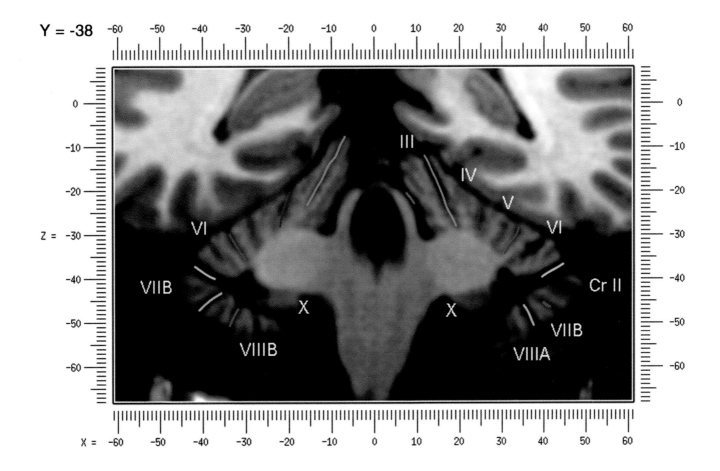

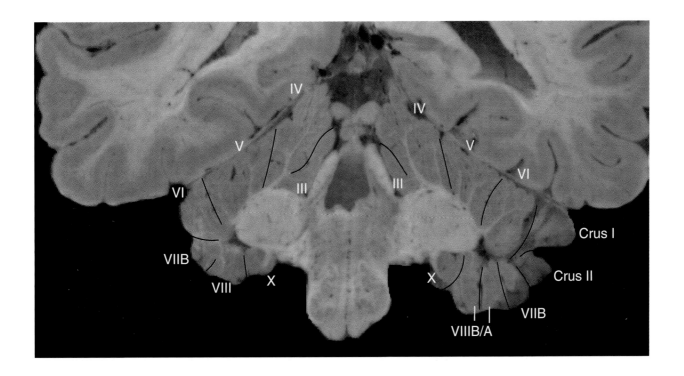

Y = -40

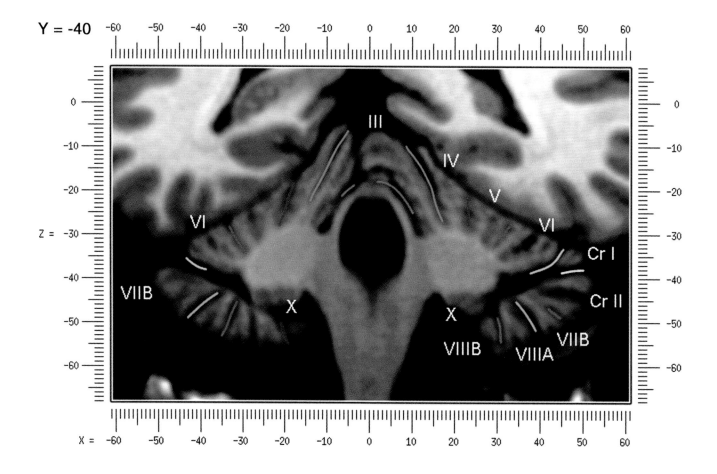

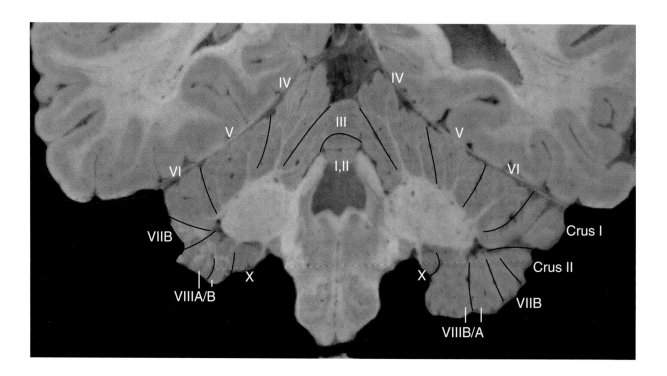

Y = -42

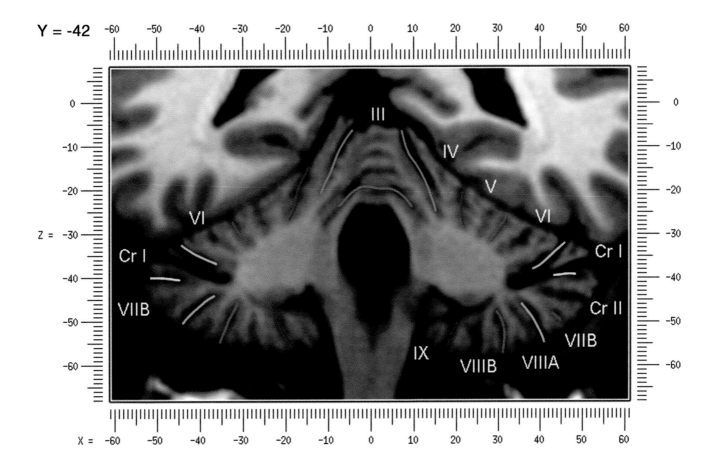

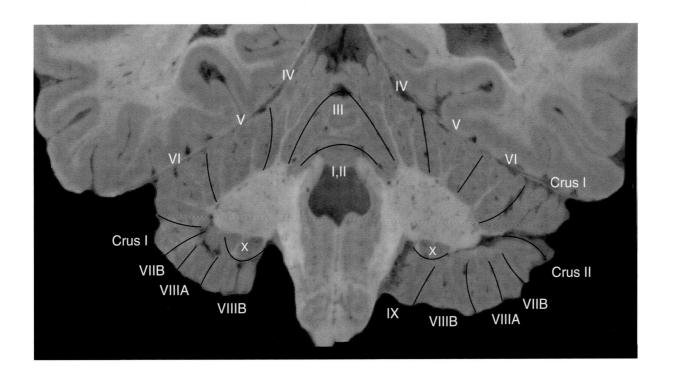

Y = -44

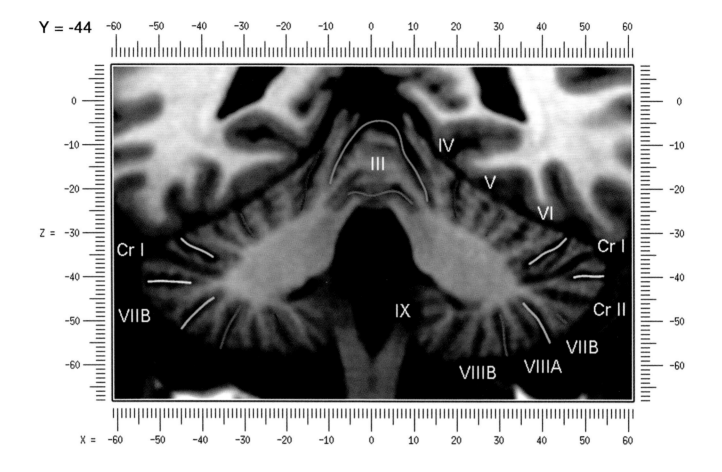

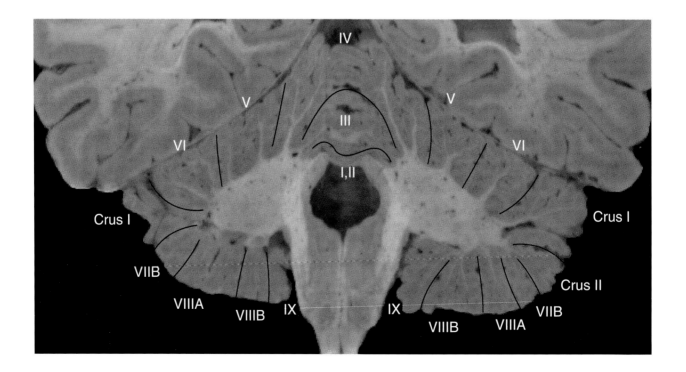

Y = -46

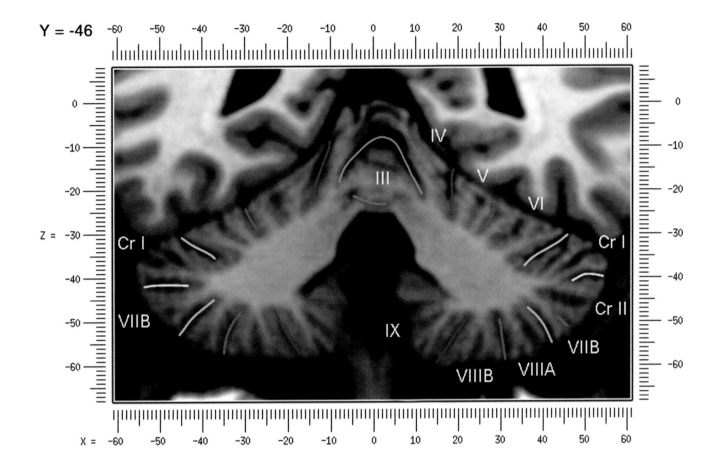

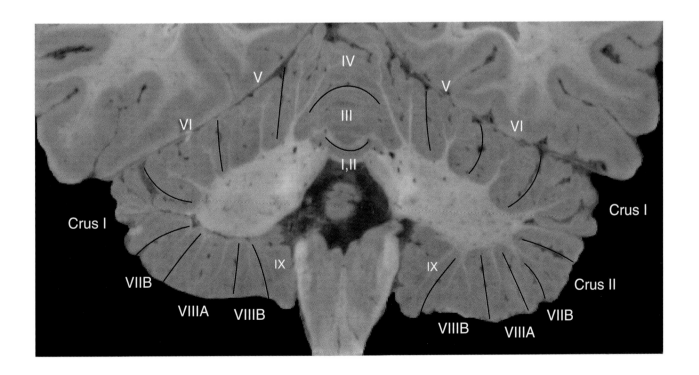

Y = -48

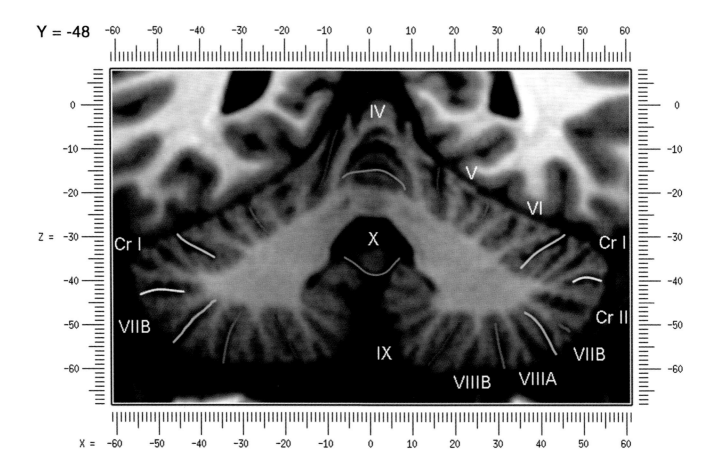

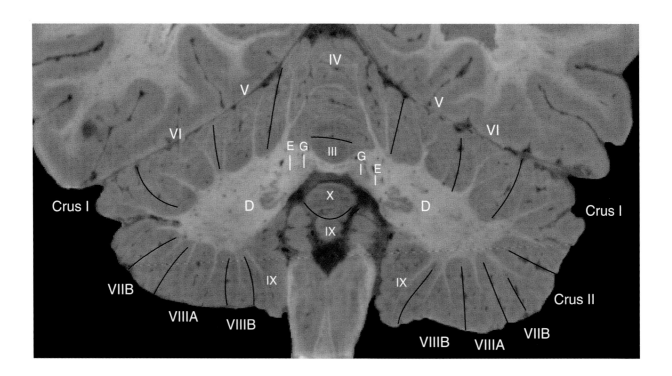

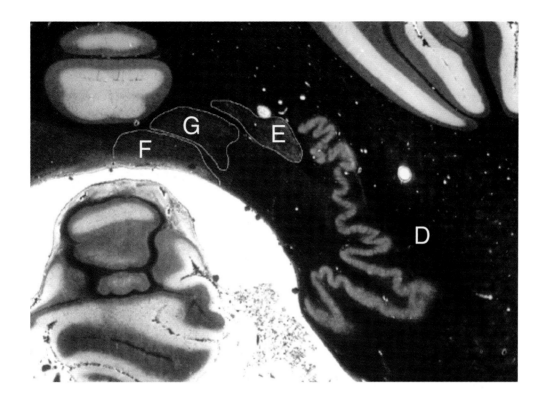

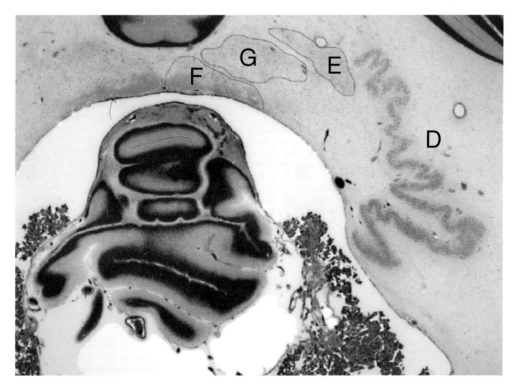

Top: Myelin stain.   Bottom: Nissl stain.
D= Dentate Nucleus, E= Emboliform Nucleus, F=Fastigial Nucleus, G= Globose Nucleus

Y = -50

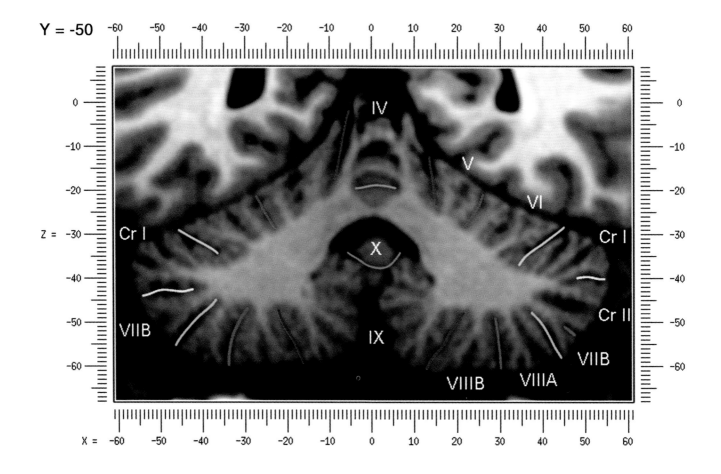

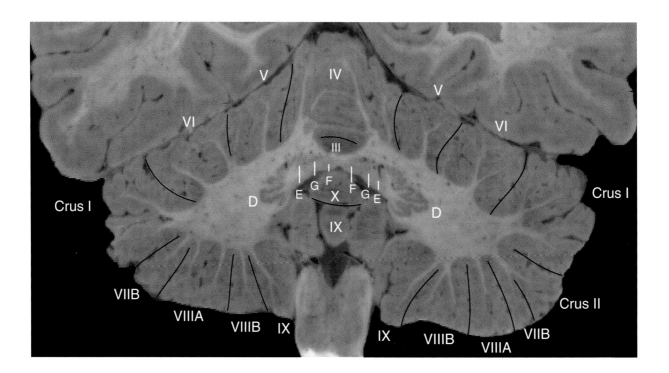

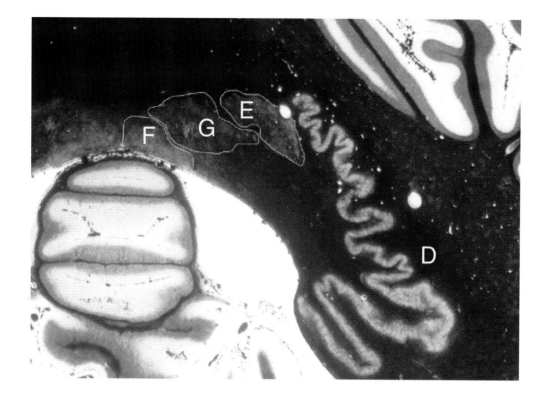

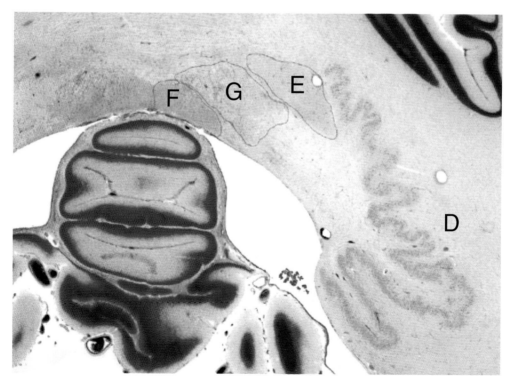

Top: Myelin stain. Bottom: Nissl stain.
D= Dentate Nucleus, E= Emboliform Nucleus, F=Fastigial Nucleus, G= Globose Nucleus

Y = -52

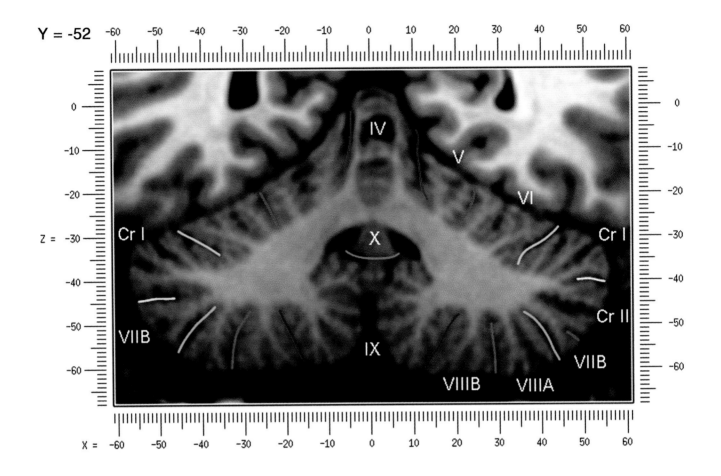

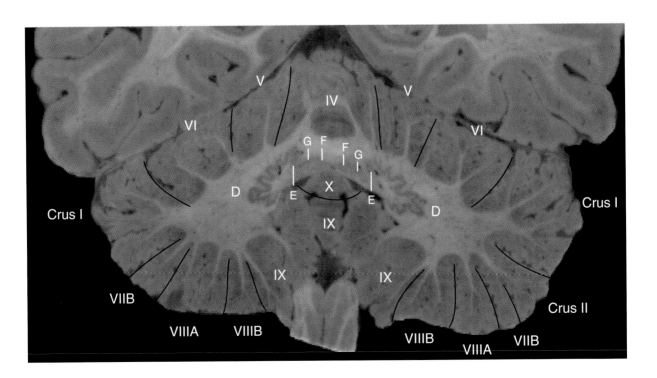

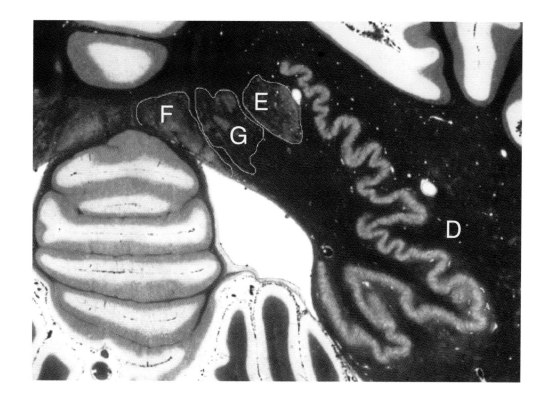

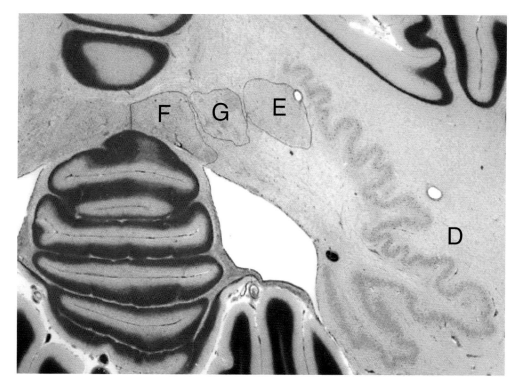

Top: Myelin stain.   Bottom: Nissl stain.
D= Dentate Nucleus, E= Emboliform Nucleus, F=Fastigial Nucleus, G= Globose Nucleus

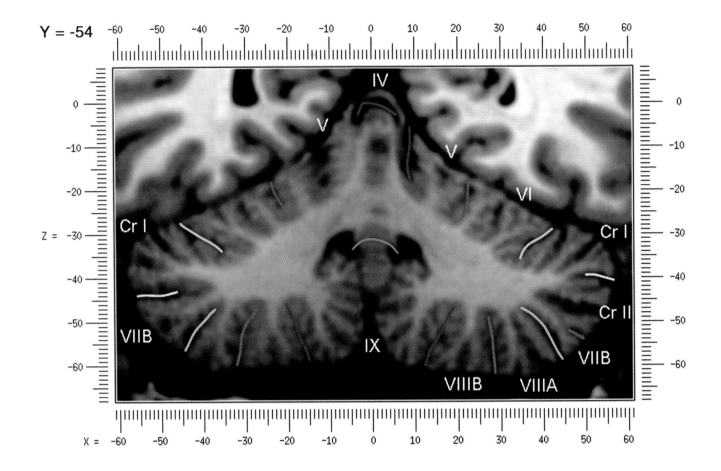

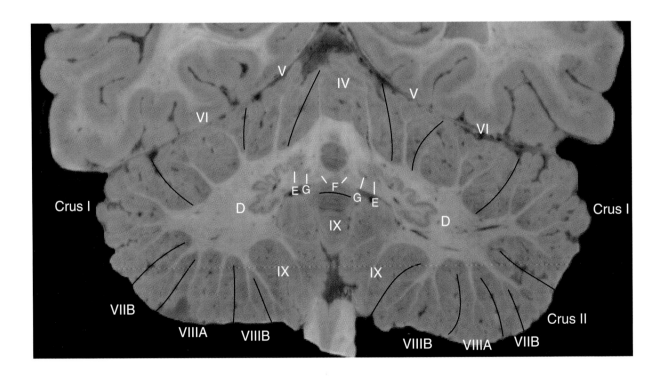

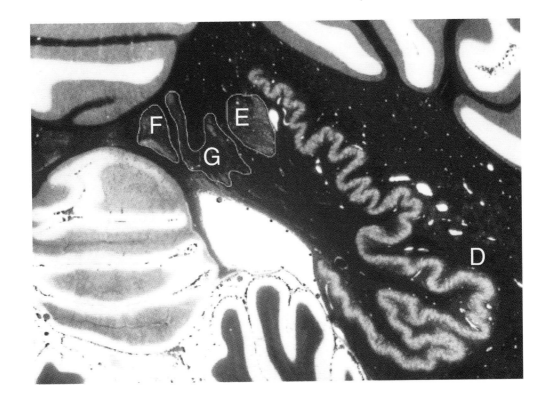

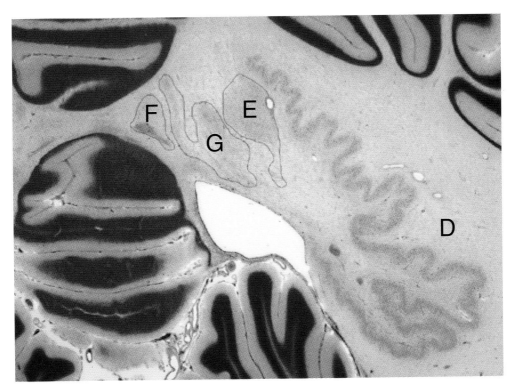

Top: Myelin stain.   Bottom: Nissl stain.
D= Dentate Nucleus, E= Emboliform Nucleus, F=Fastigial Nucleus, G= Globose Nucleus

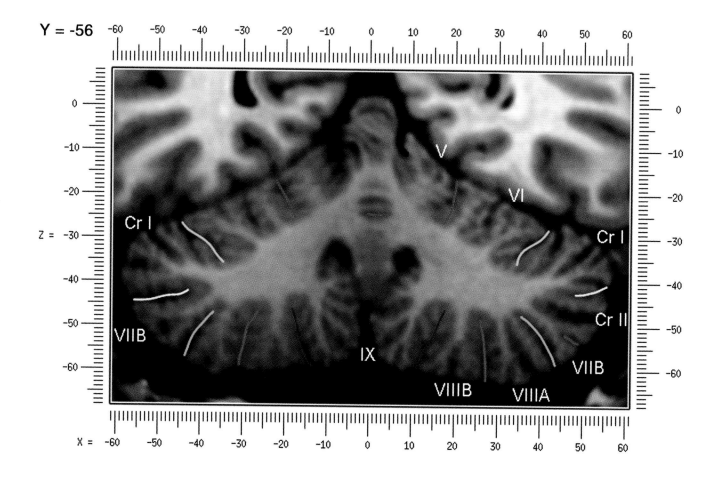

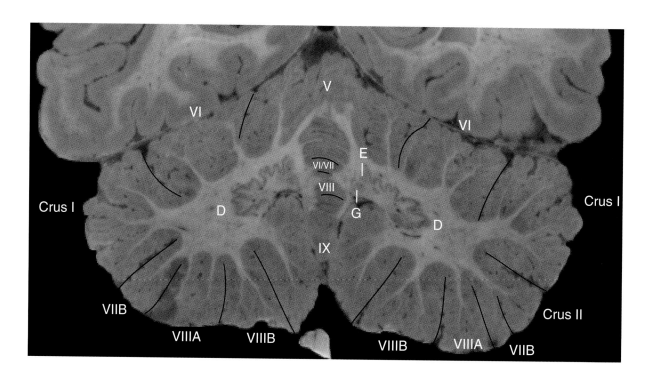

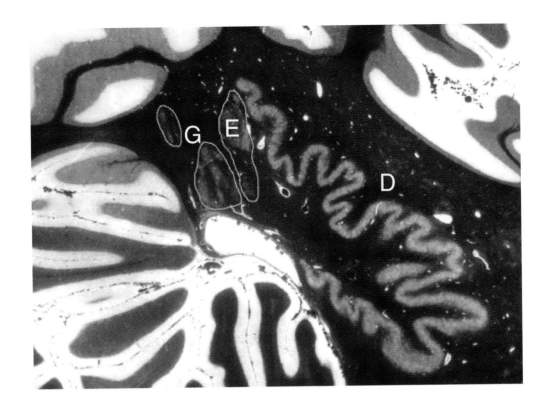

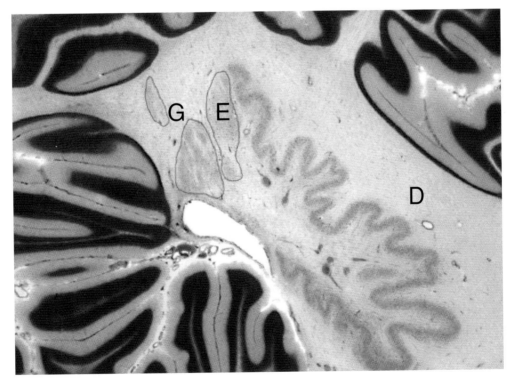

Top: Myelin stain.   Bottom: Nissl stain.
D= Dentate Nucleus, E= Emboliform Nucleus, G= Globose Nucleus

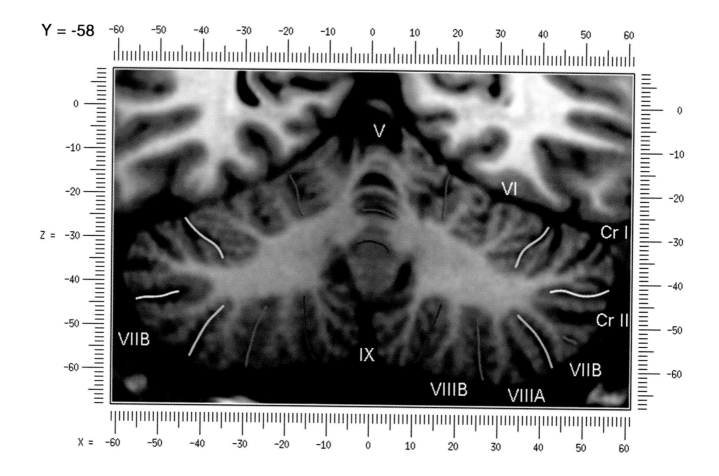

Y = -58

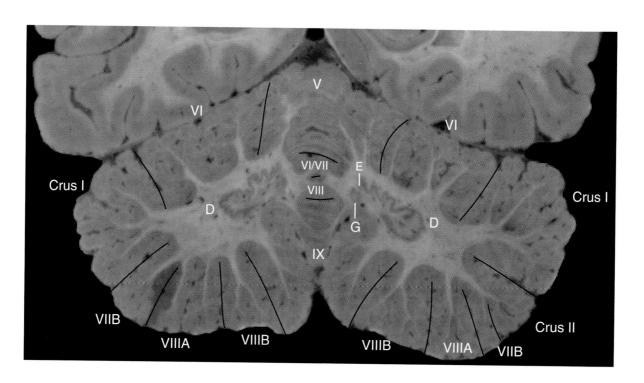

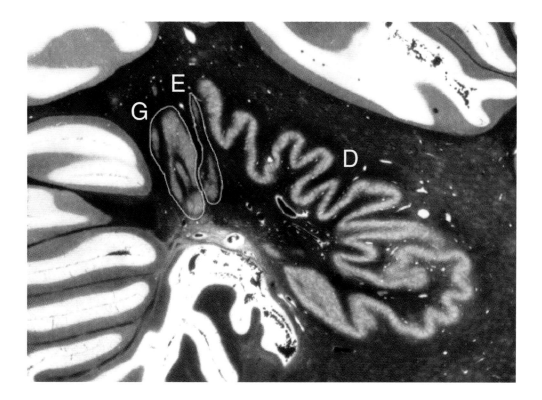

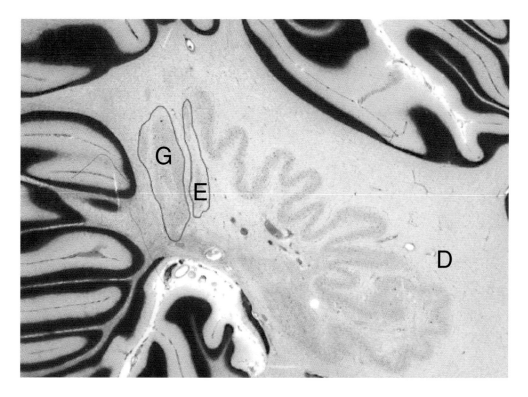

Top: Myelin stain.   Bottom: Nissl stain.
D= Dentate Nucleus, E= Emboliform Nucleus, G= Globose Nucleus

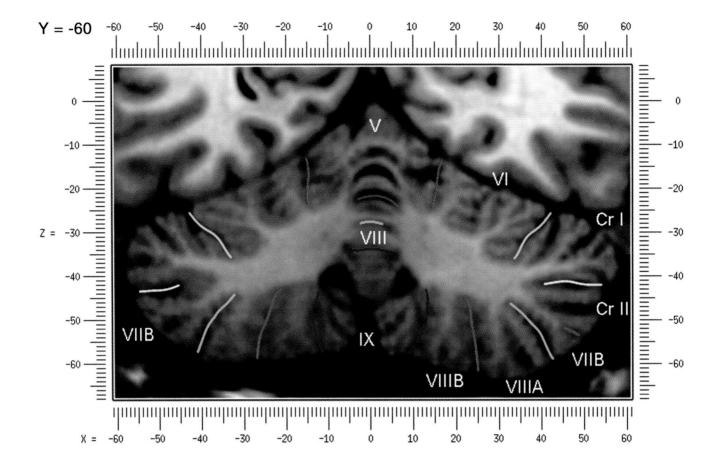

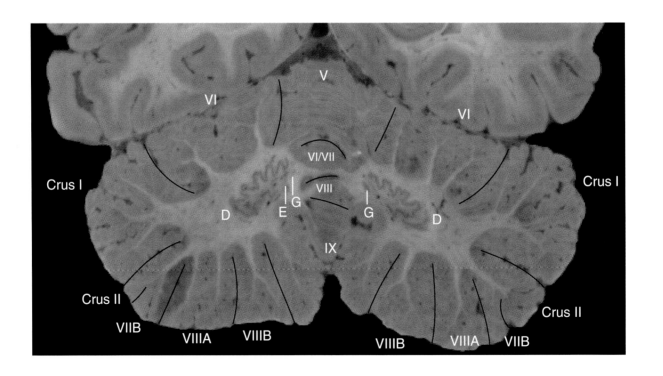

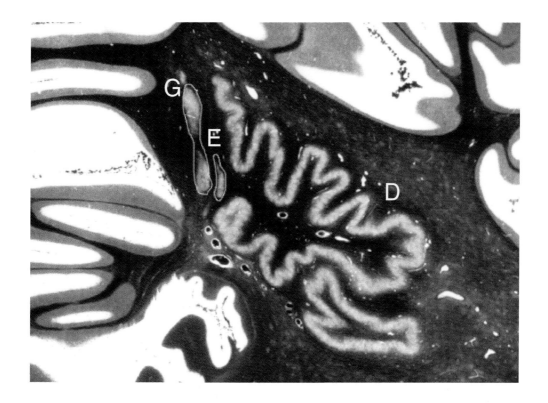

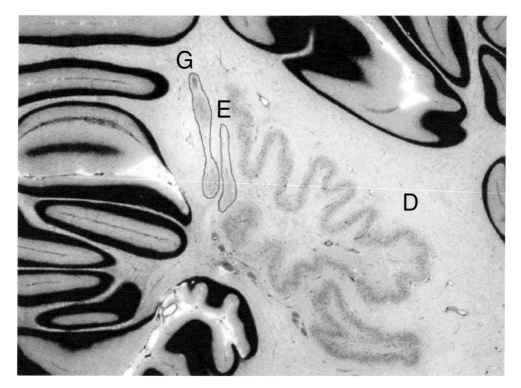

Top: Myelin stain.   Bottom: Nissl stain.
D= Dentate Nucleus, E= Emboliform Nucleus, G= Globose Nucleus

Y = -62

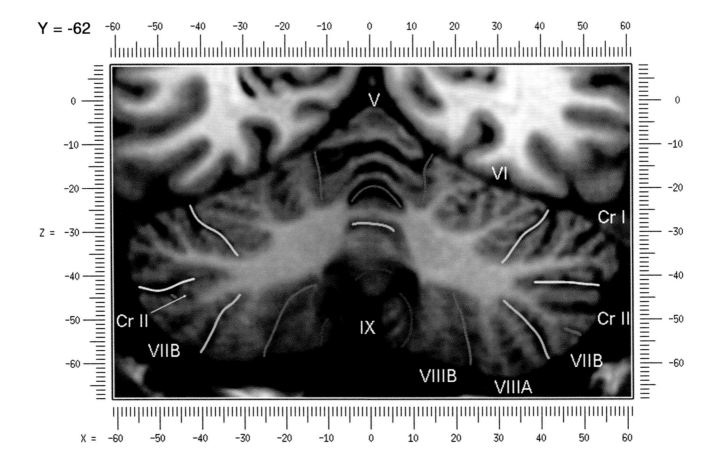

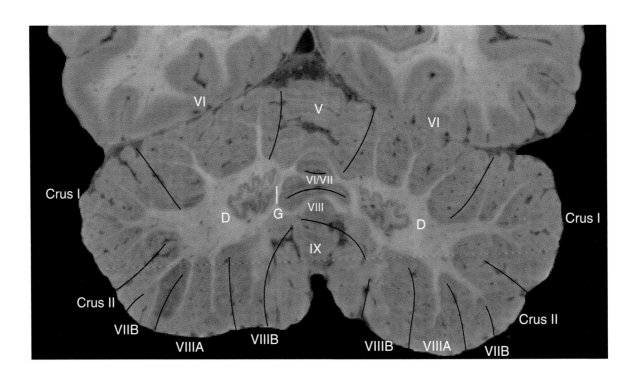

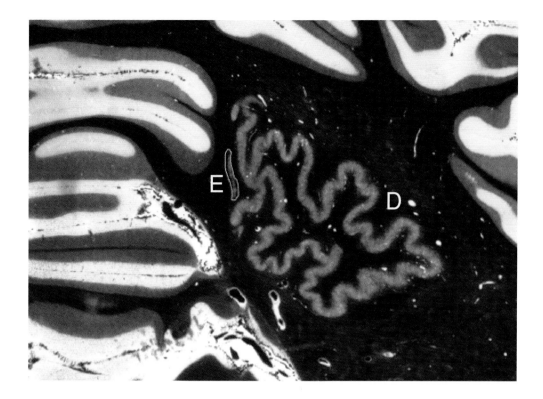

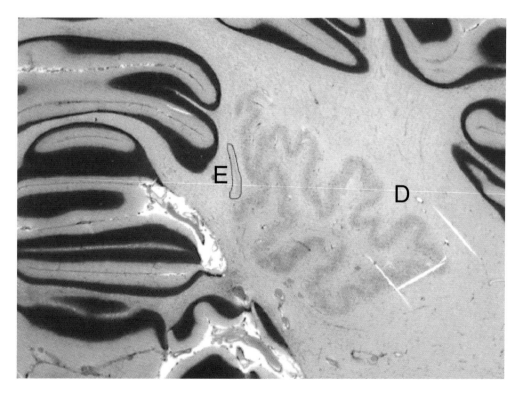

Top: Myelin stain.   Bottom: Nissl stain.
D= Dentate Nucleus, E= Emboliform Nucleus

Y = -64

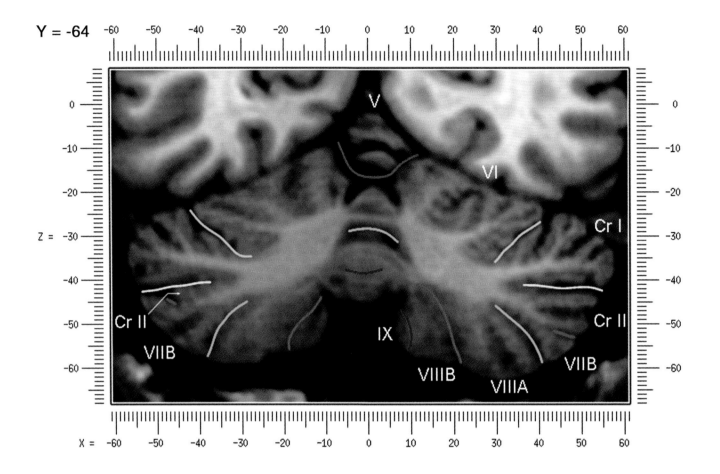

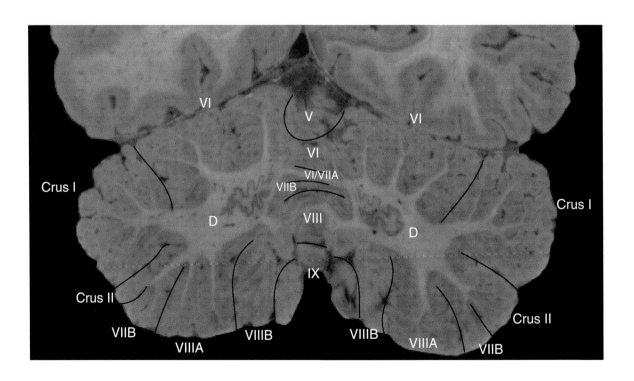

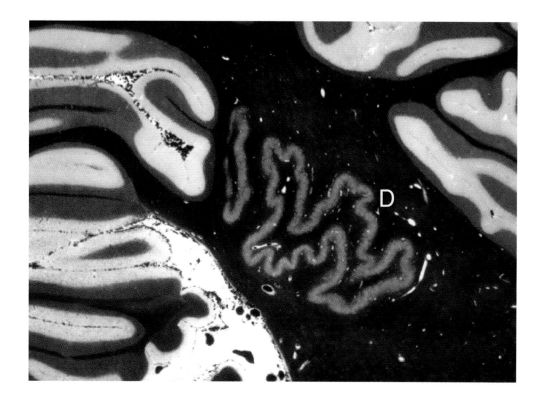

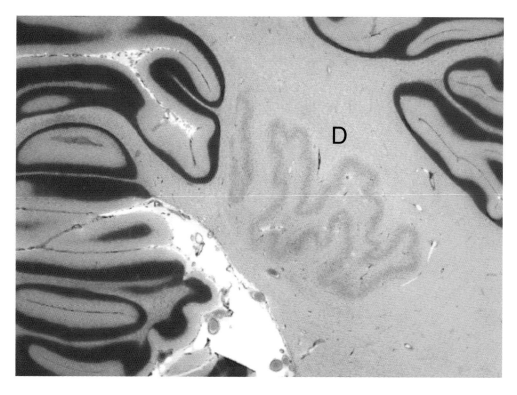

Top: Myelin stain.   Bottom: Nissl stain.
D= Dentate Nucleus

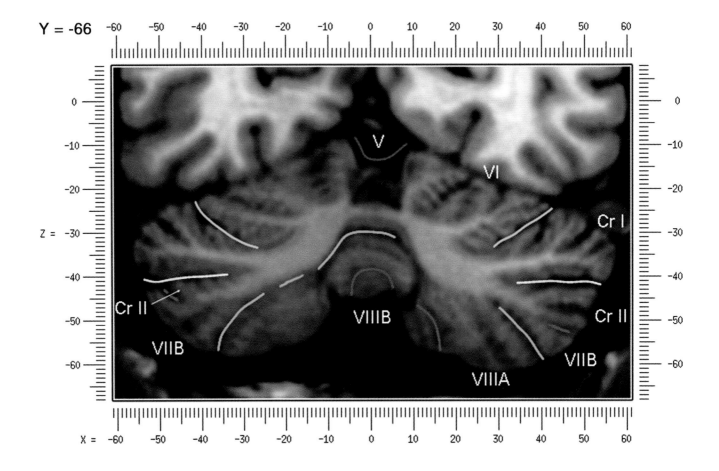

Y = -66

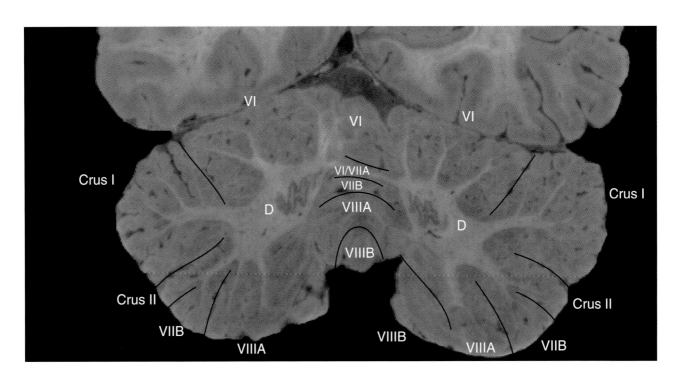

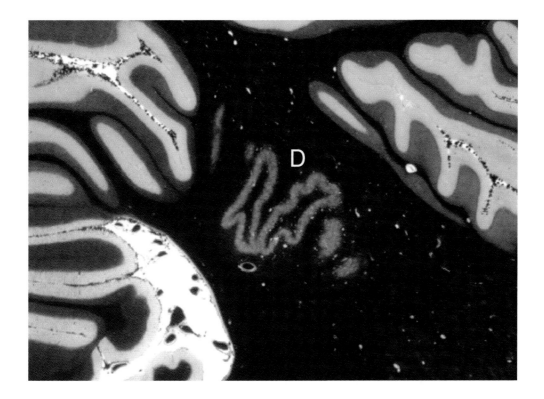

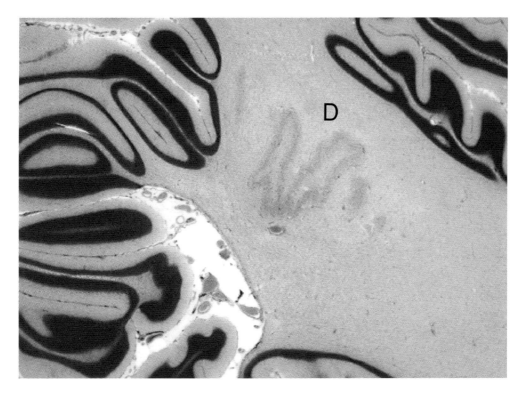

Top: Myelin stain.   Bottom: Nissl stain.
D= Dentate Nucleus

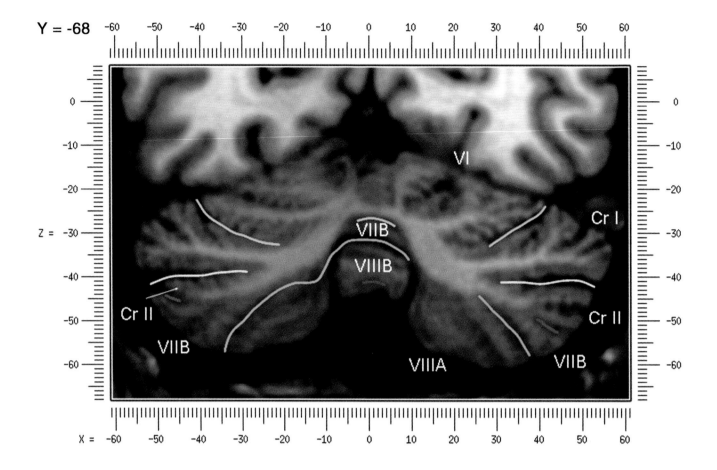

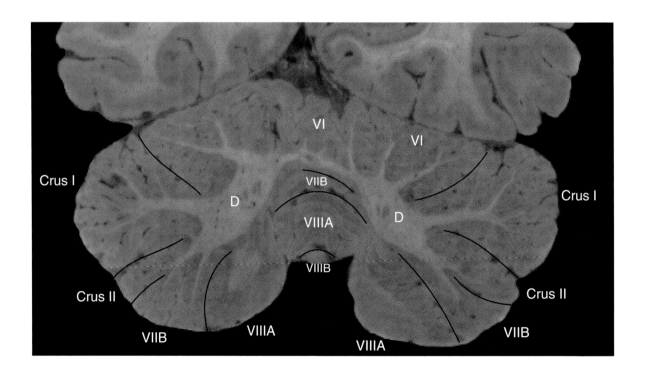

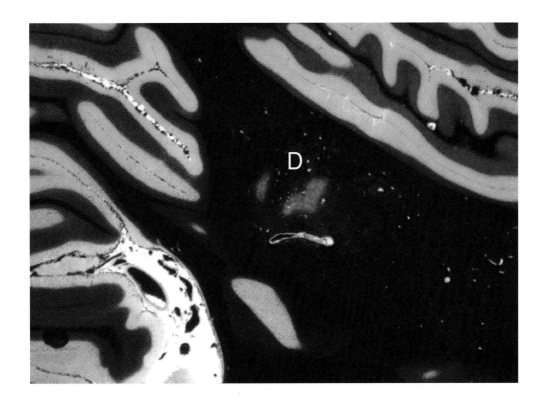

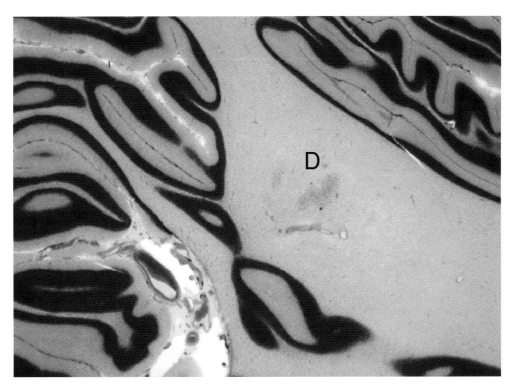

Top: Myelin stain.   Bottom: Nissl stain.
D= Dentate Nucleus

Y = -70

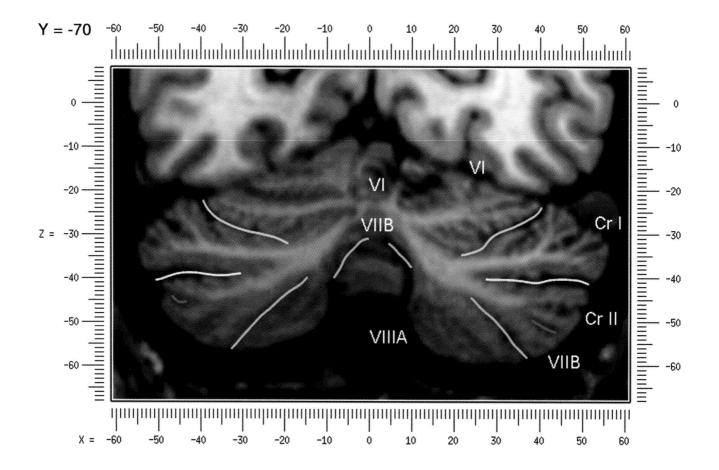

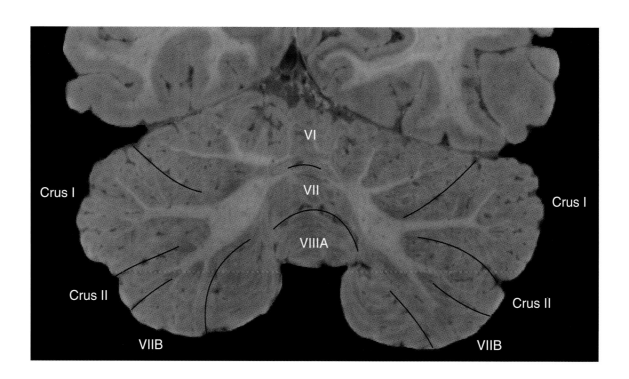

Y = -72

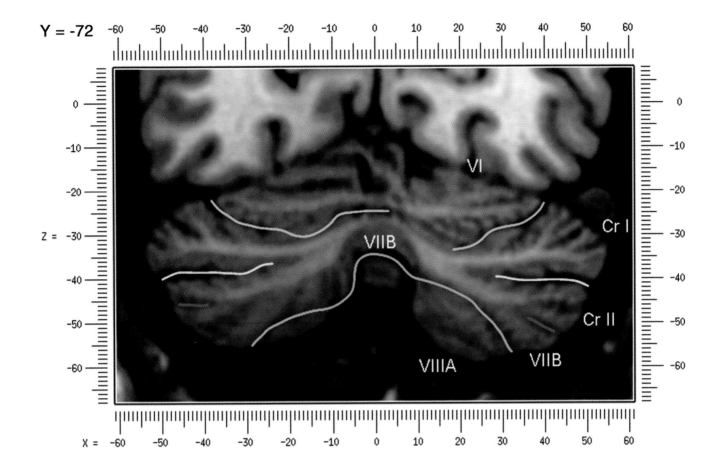

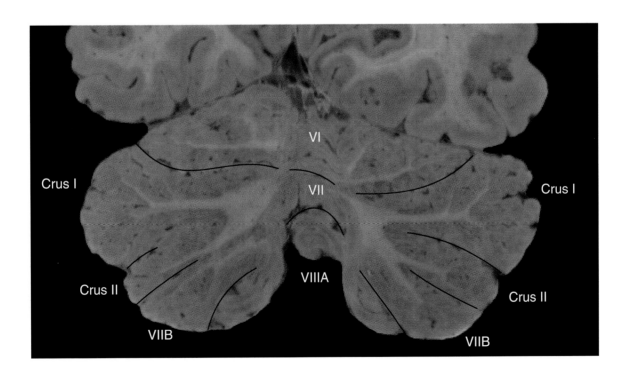

Y = -74

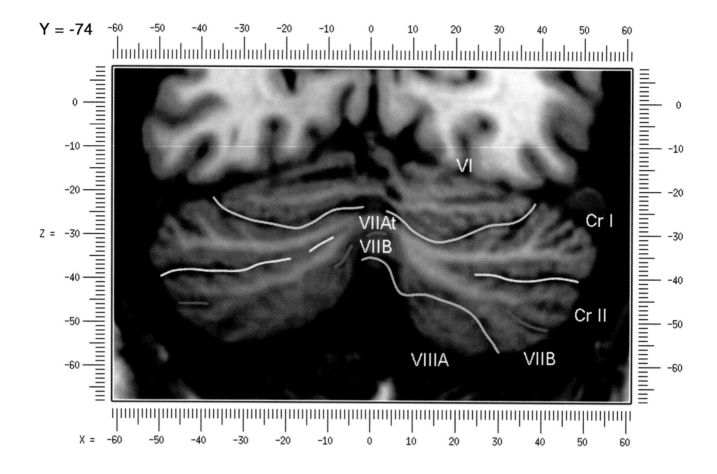

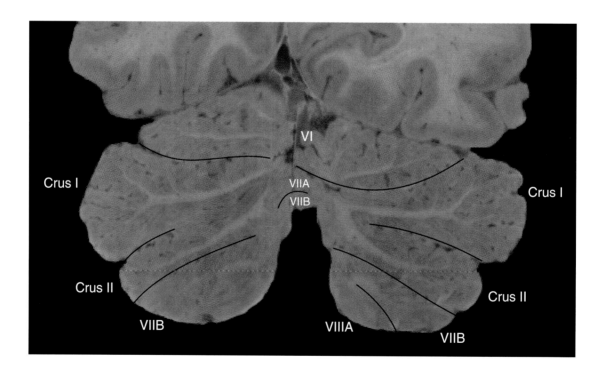

Y = -76

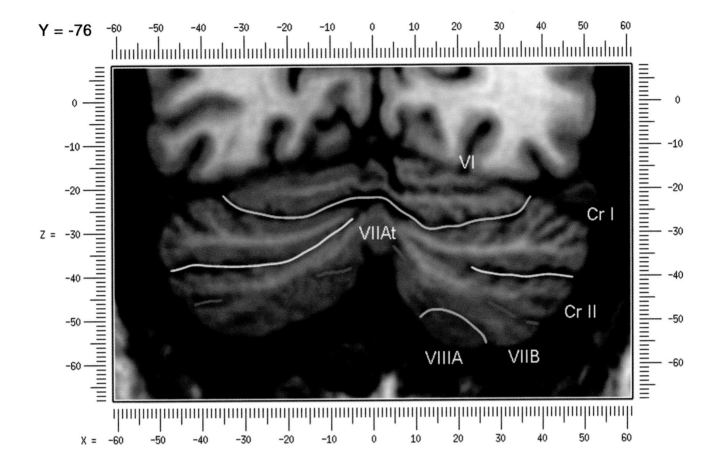

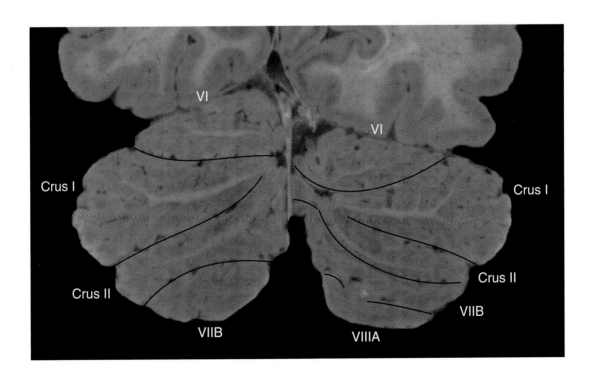

Y = -78

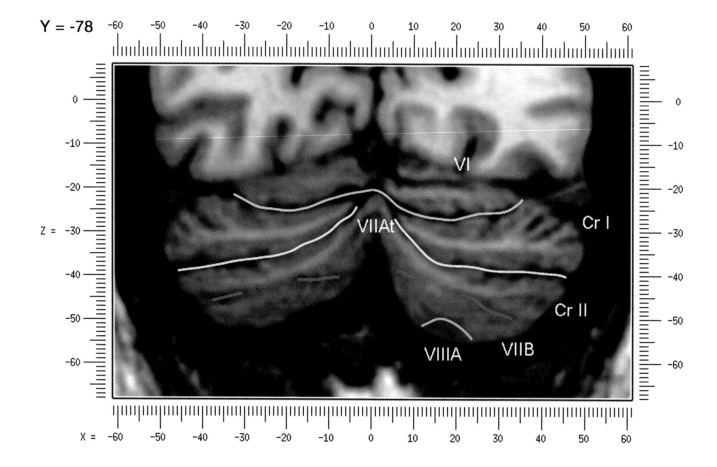

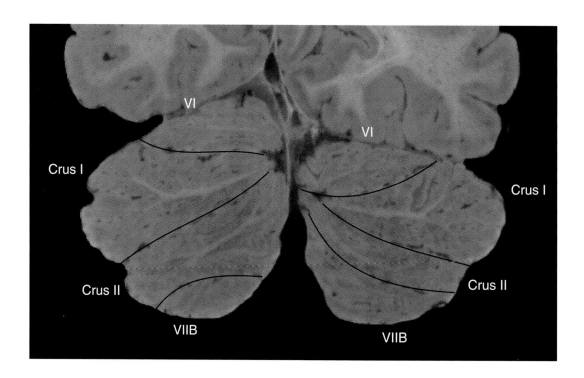

Y = -80

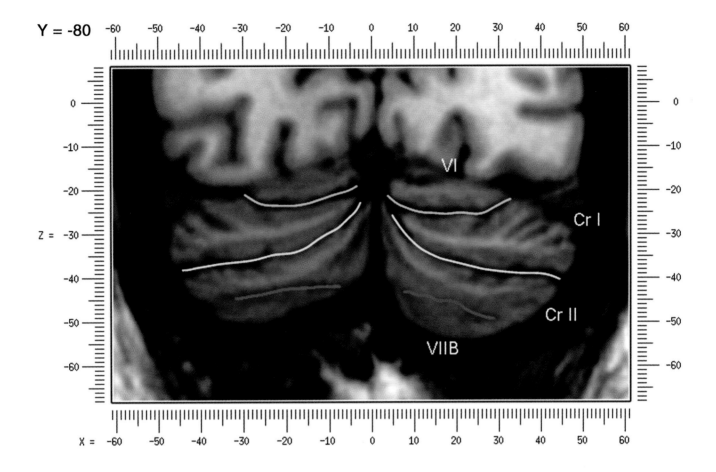

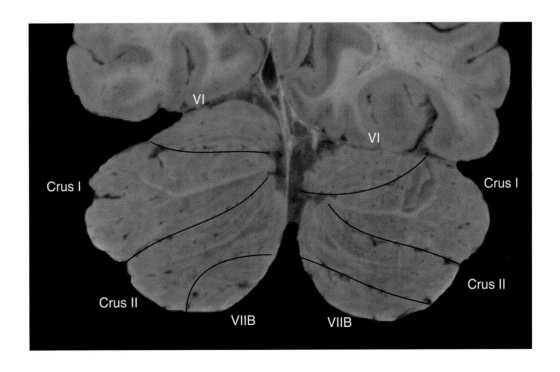

Y = -82

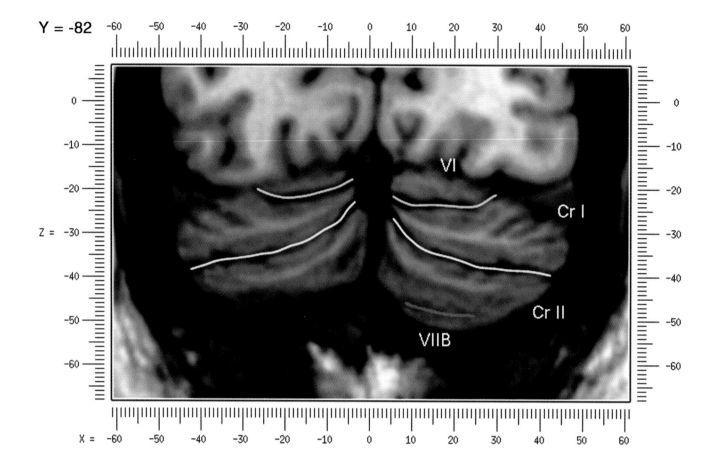

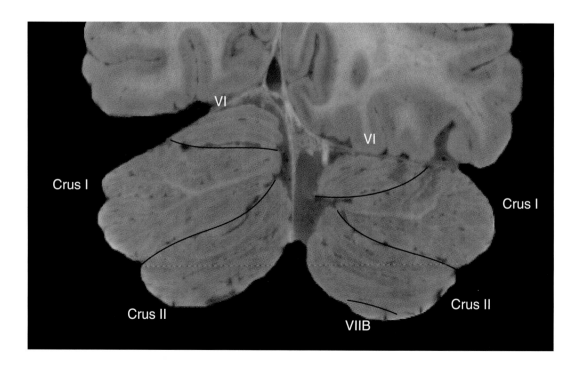

Y = -84

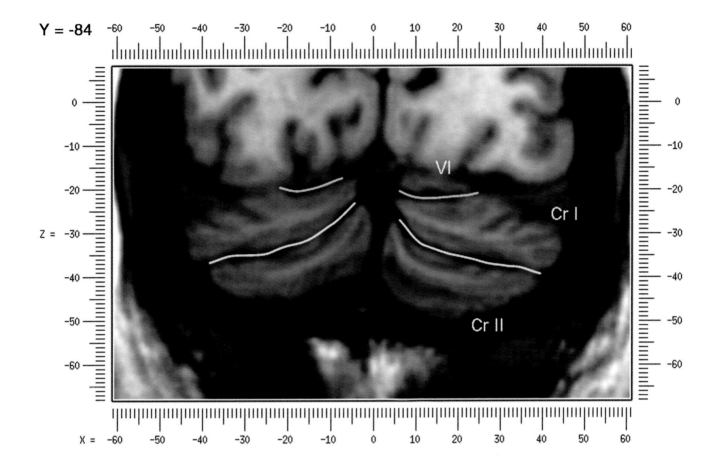

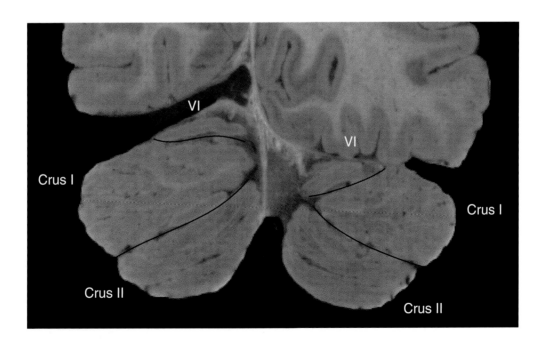

Y = -86

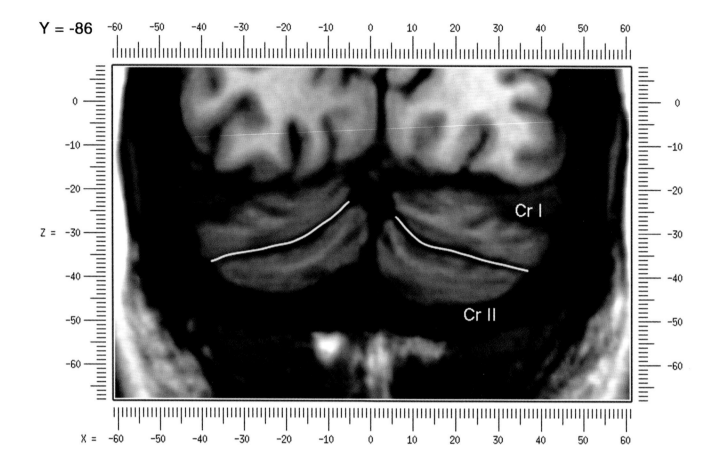

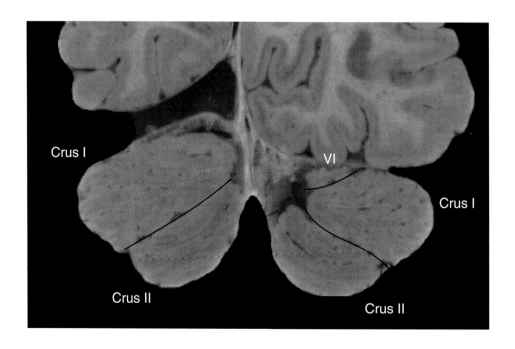

Y = -88

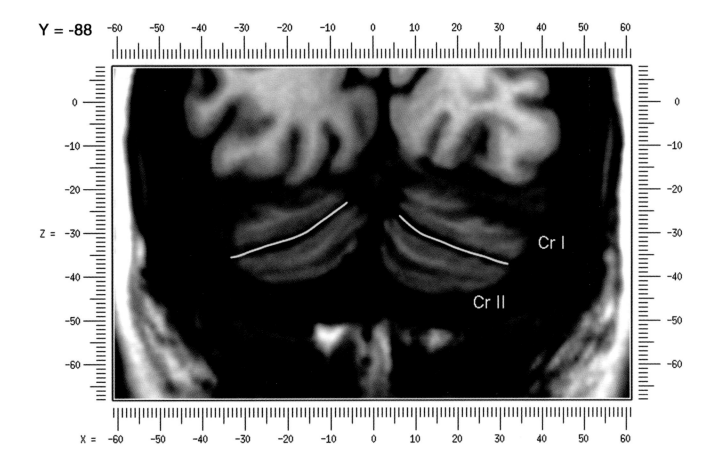

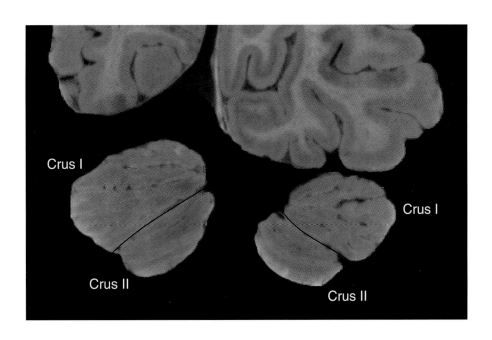

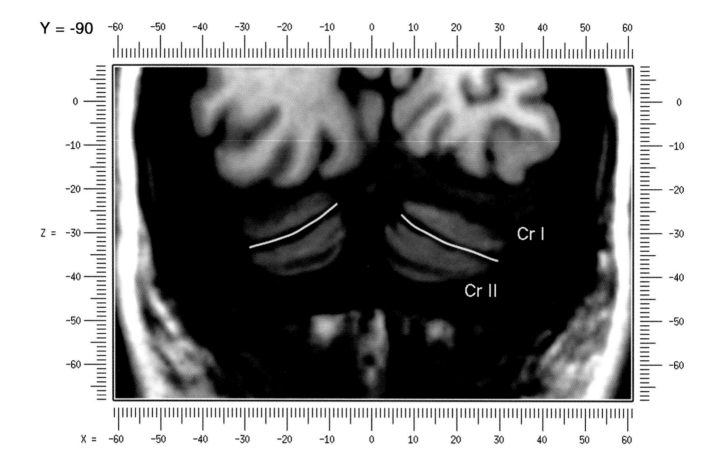

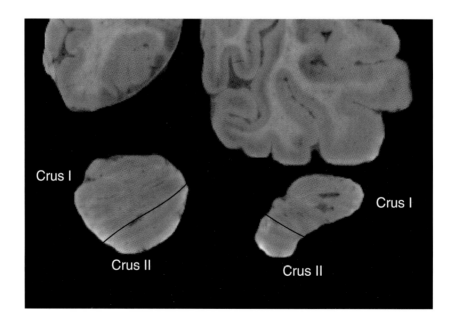

Horizontal Series

Z = -9

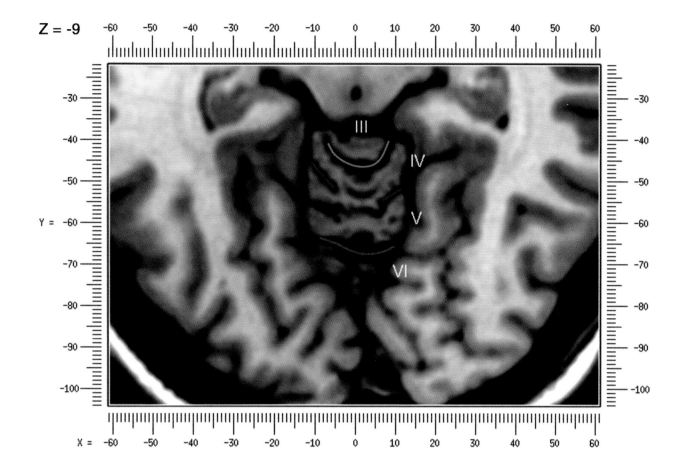

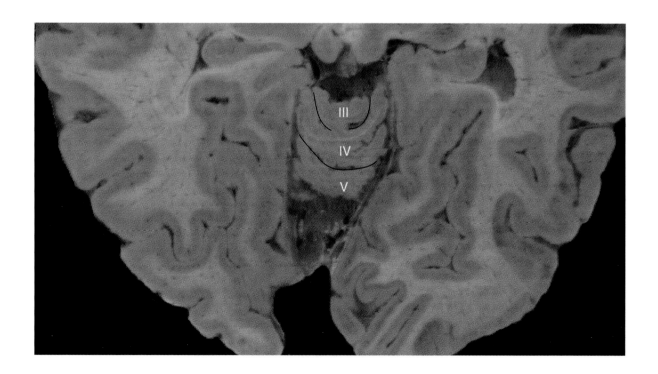

Z = -11

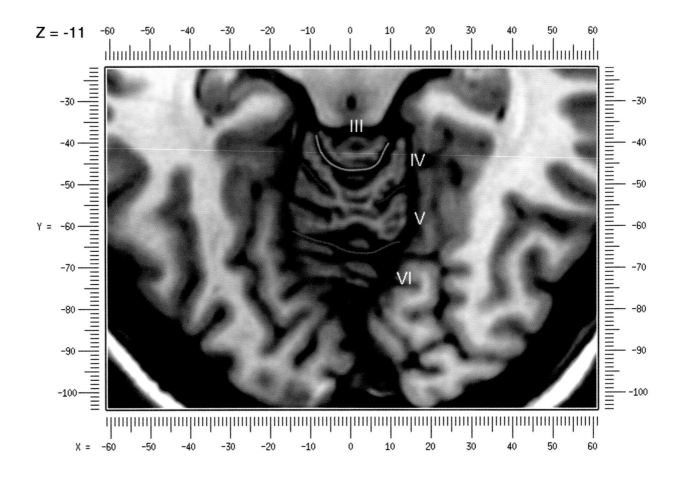

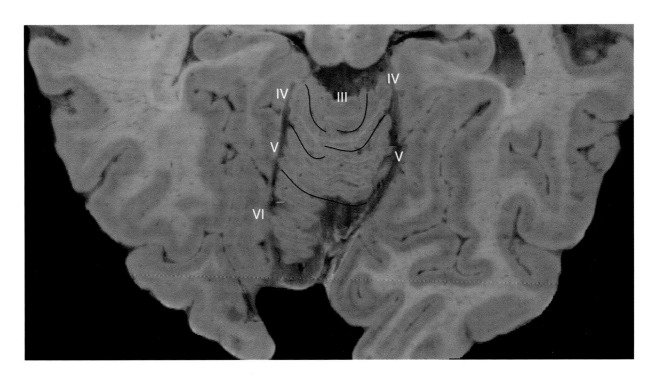

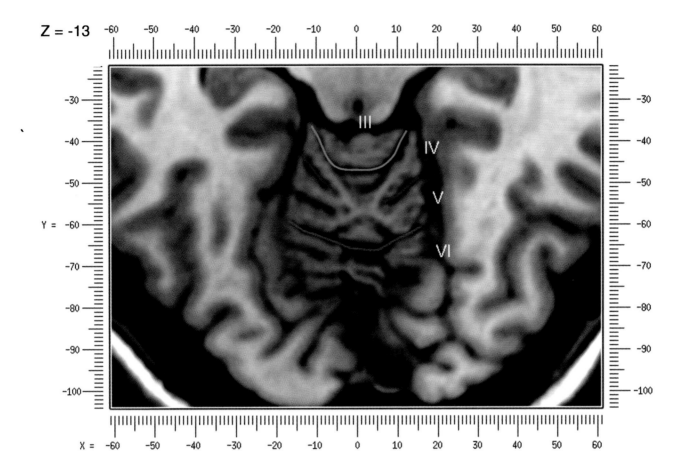

Z = -13

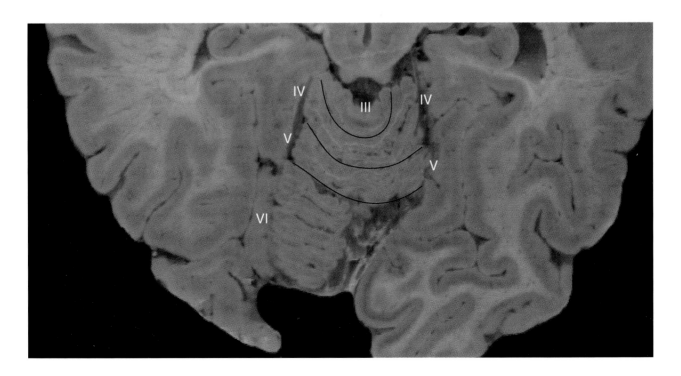

Z = -15

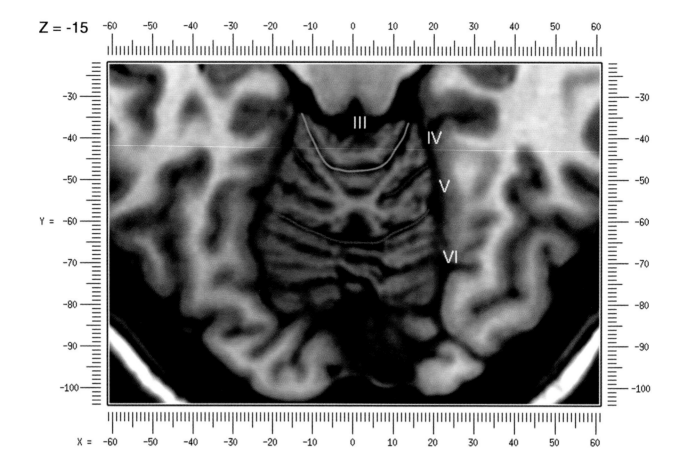

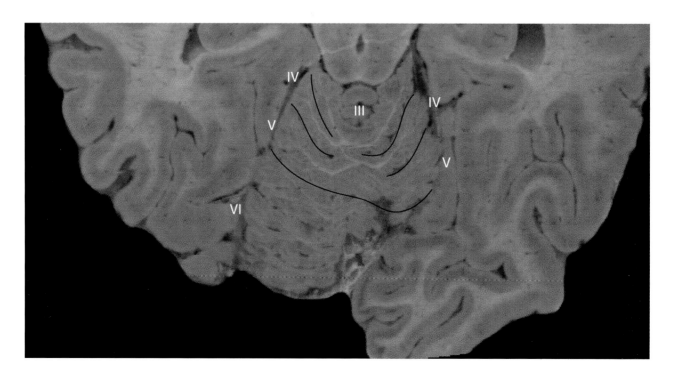

Z = -17

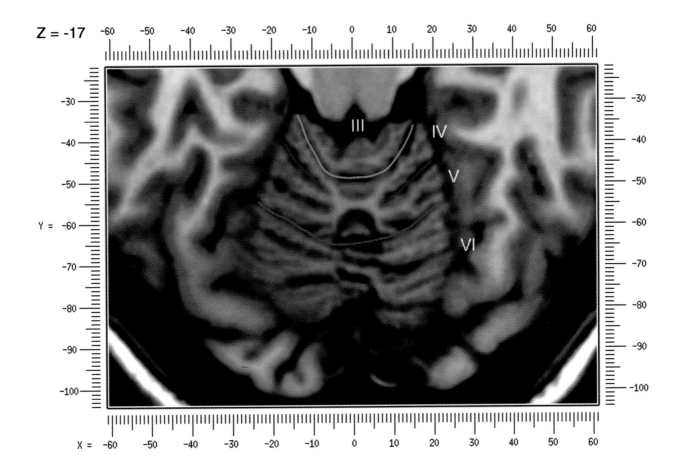

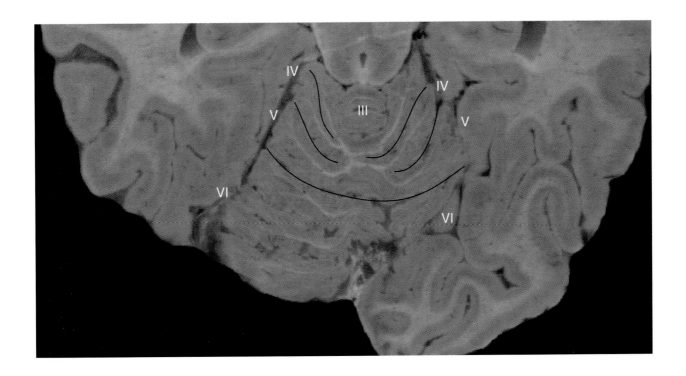

Z = -19

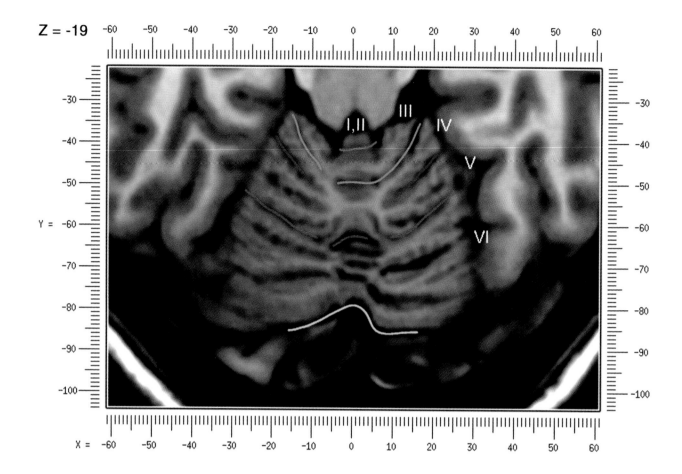

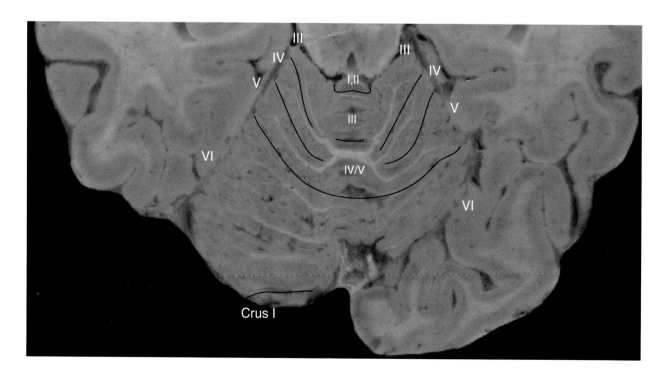

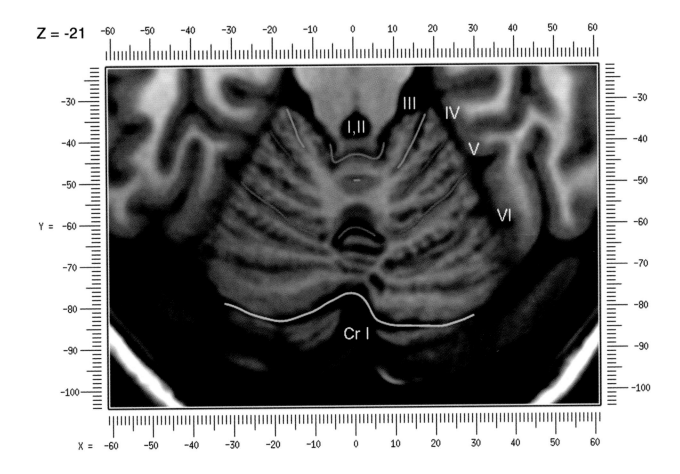

Z = -21

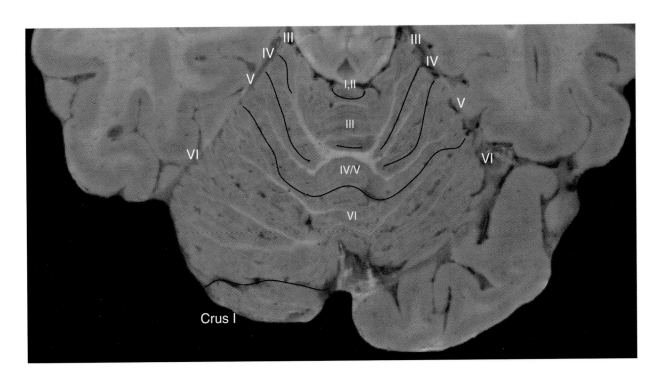

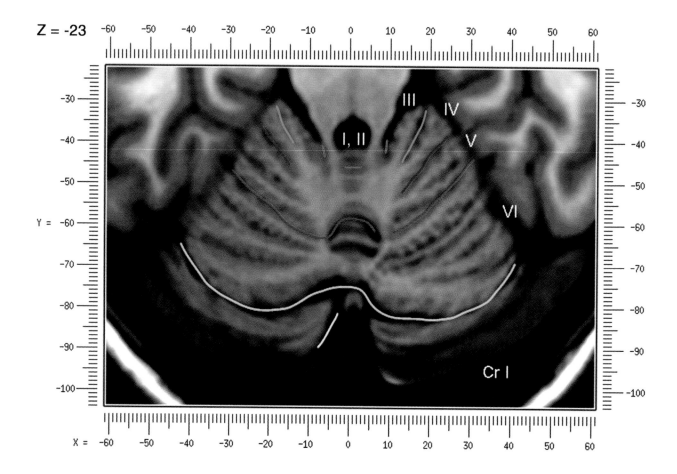

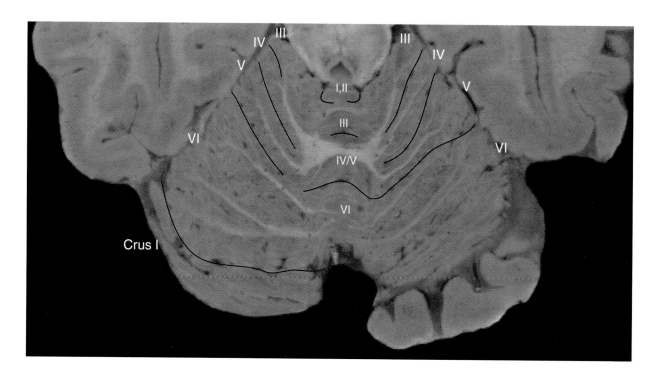

Z = -25

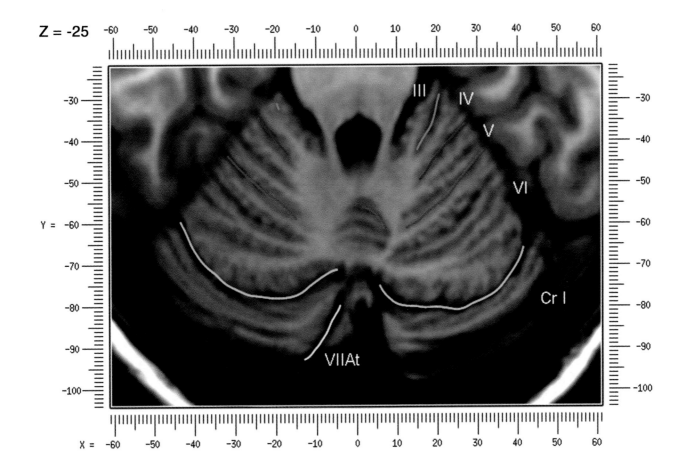

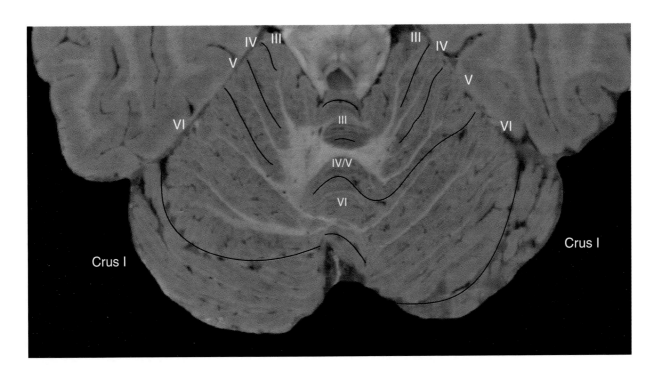

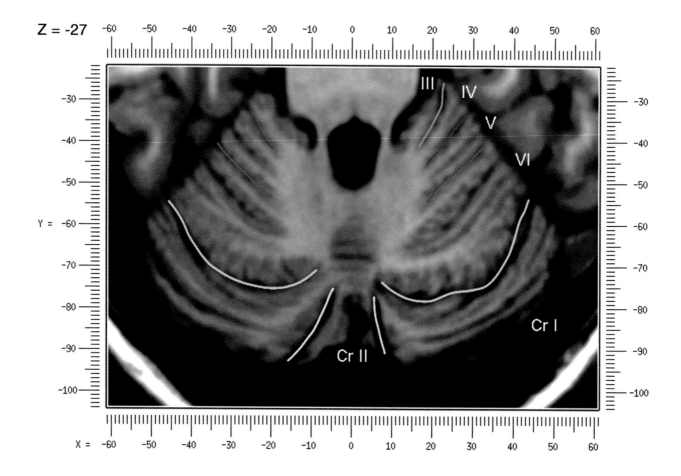

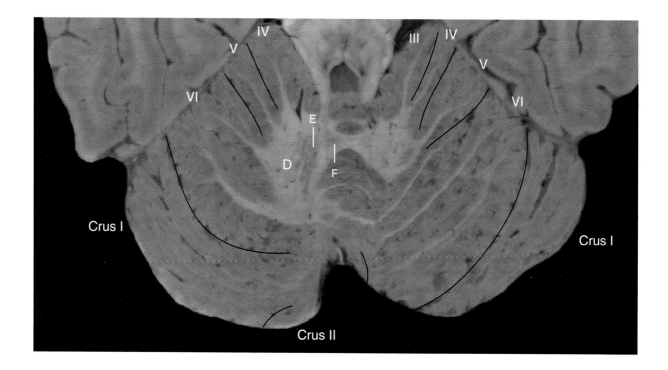

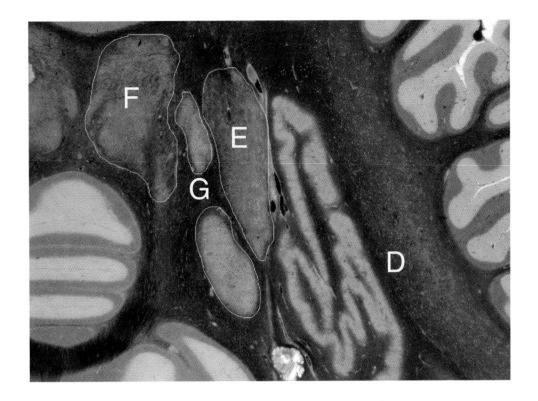

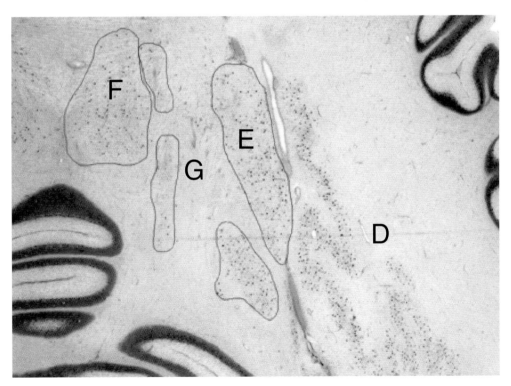

Top: Myelin stain.   Bottom: Nissl stain.
D= Dentate Nucleus, E= Emboliform Nucleus, F=Fastigial Nucleus, G= Globose Nucleus

Z = -29

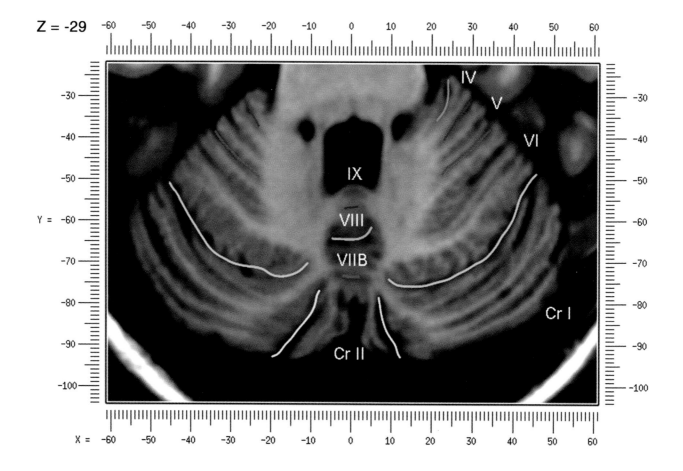

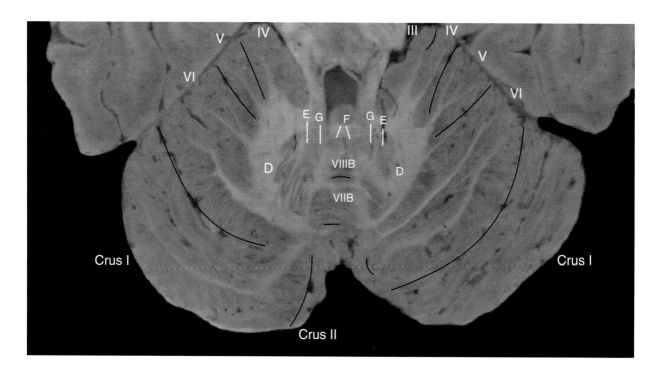

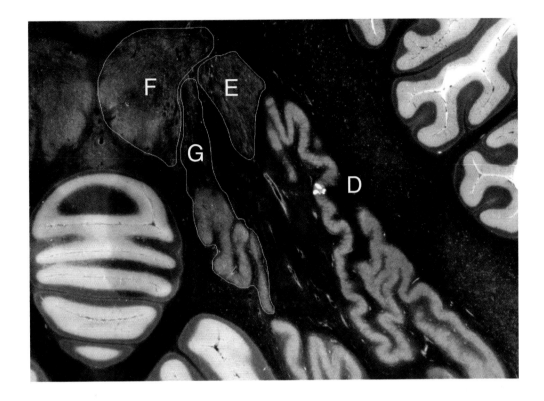

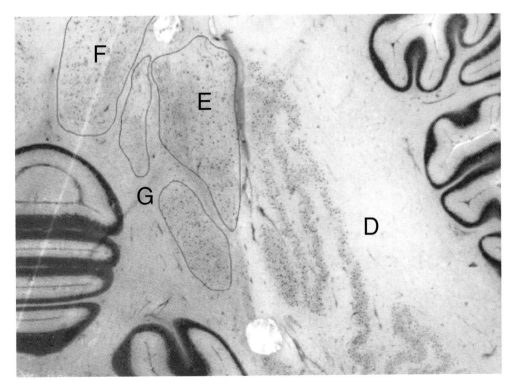

Top: Myelin stain.   Bottom: Nissl stain.
D= Dentate Nucleus, E= Emboliform Nucleus, F=Fastigial Nucleus, G= Globose Nucleus

Z = -31

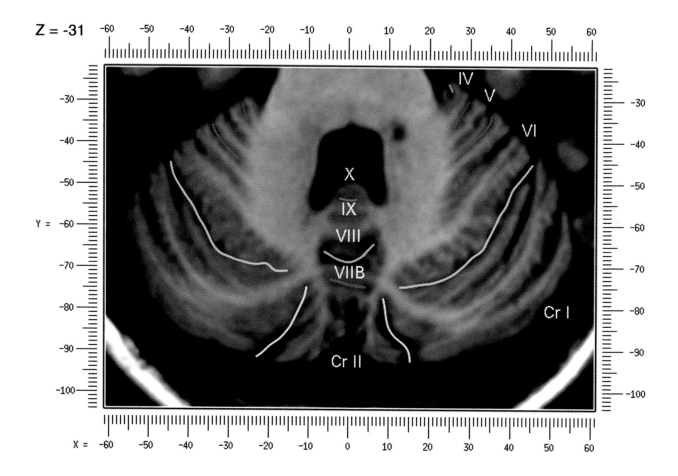

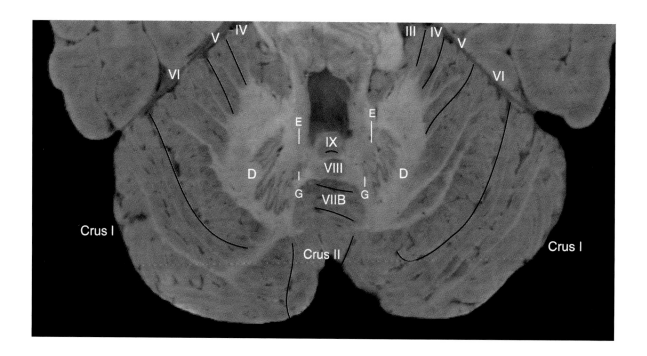

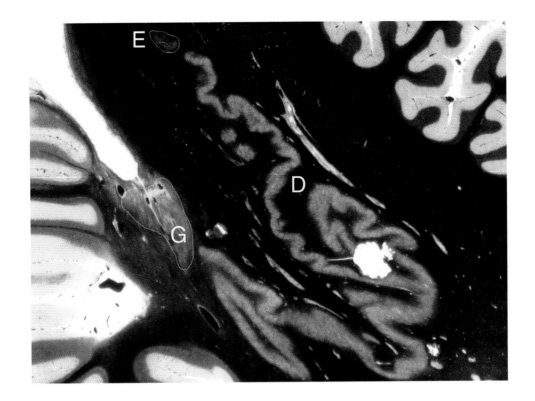

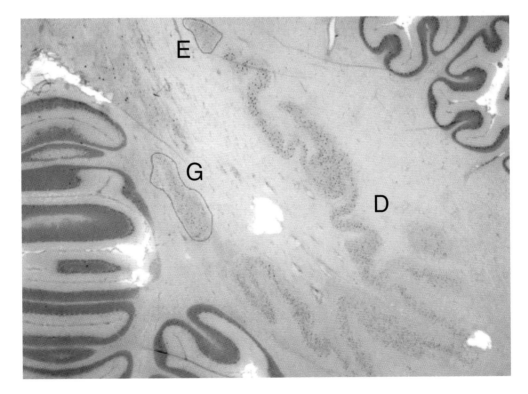

Top: Myelin stain. Bottom: Nissl stain.
D= Dentate Nucleus, E= Emboliform Nucleus, G= Globose Nucleus

Z = -33

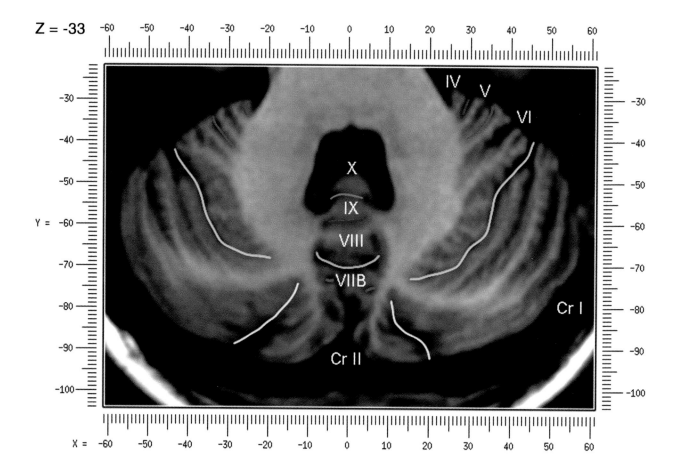

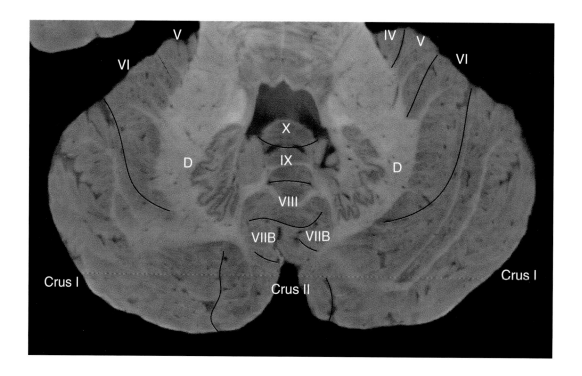

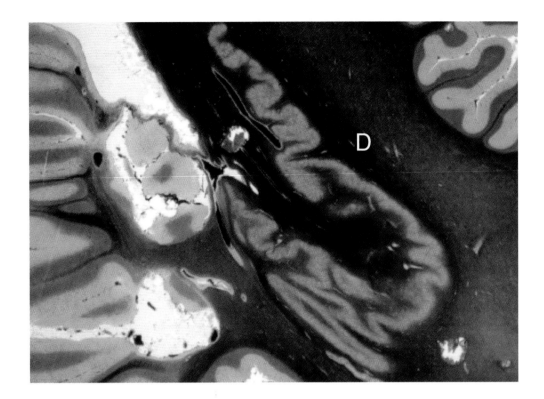

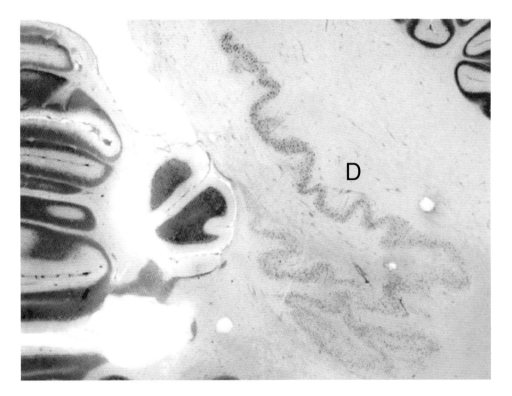

Top: Myelin stain.   Bottom: Nissl stain.
D= Dentate Nucleus

Z = -35

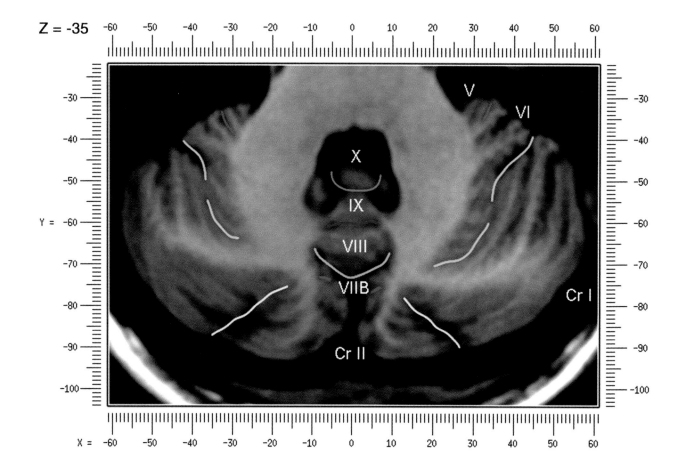

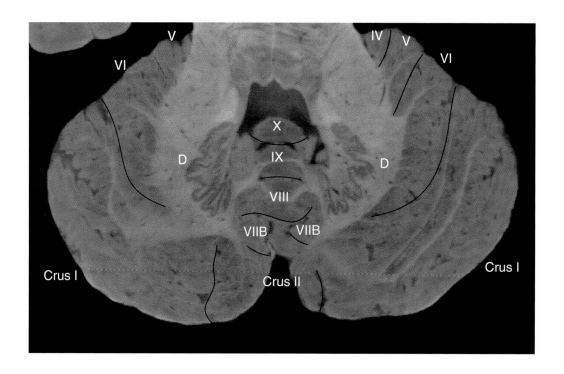

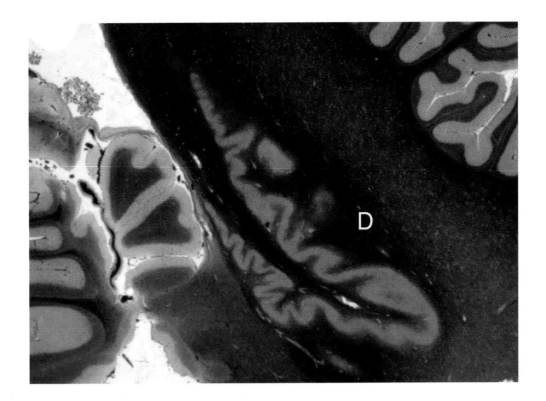

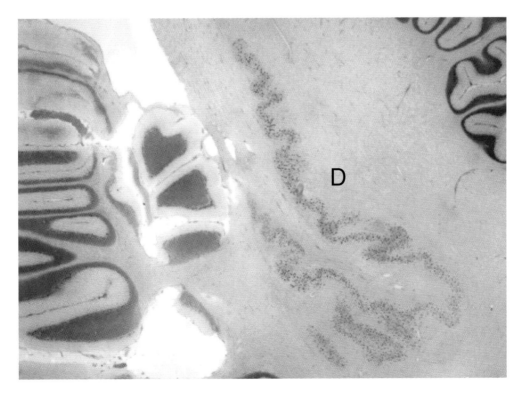

Top: Myelin stain.   Bottom: Nissl stain.
D= Dentate Nucleus

Z = -37

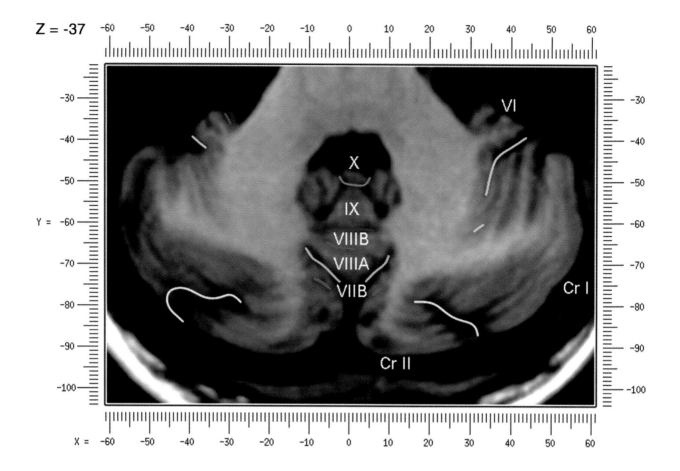

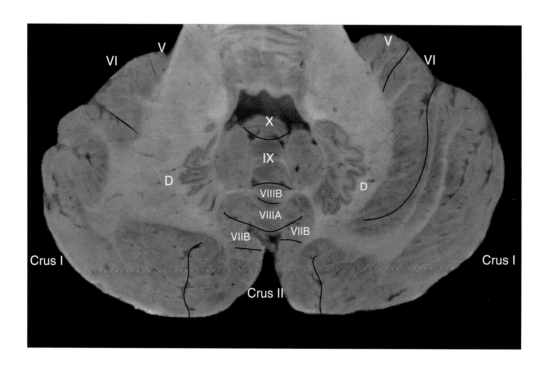

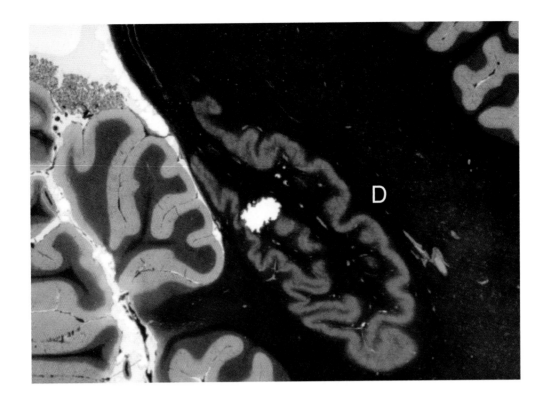

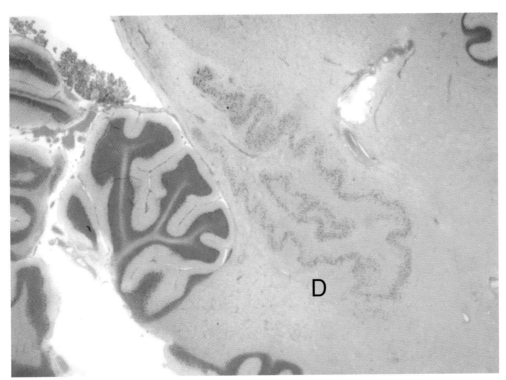

Top: Myelin stain.   Bottom: Nissl stain.
D= Dentate Nucleus

Z = -39

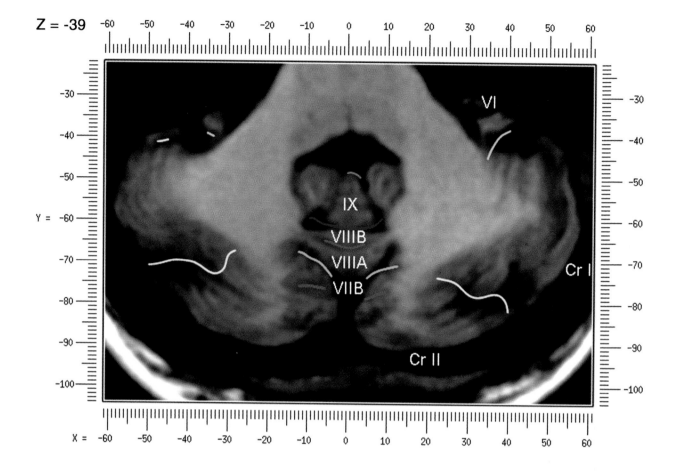

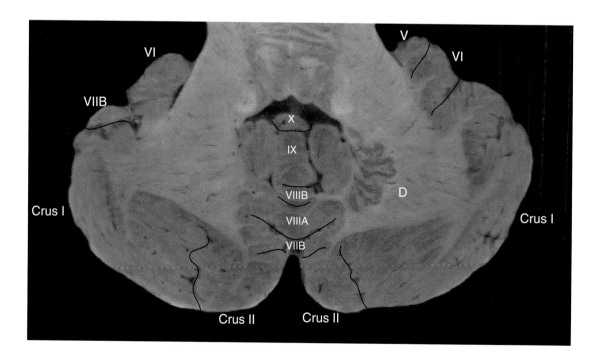

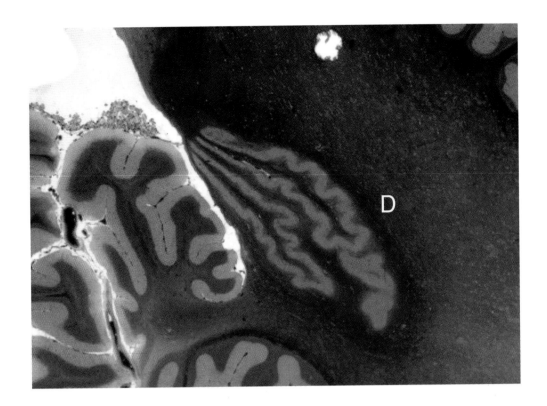

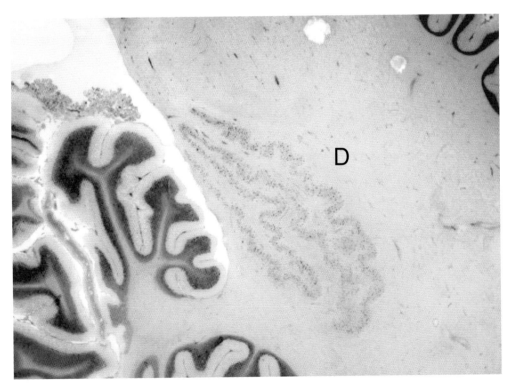

Top: Myelin stain.   Bottom: Nissl stain.
D= Dentate Nucleus

Z = -41

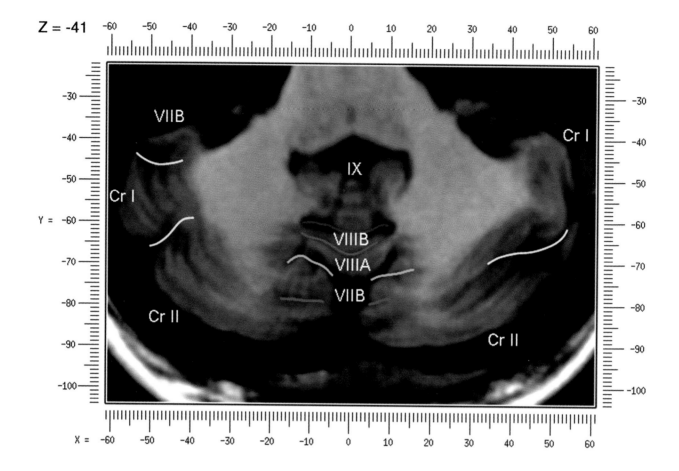

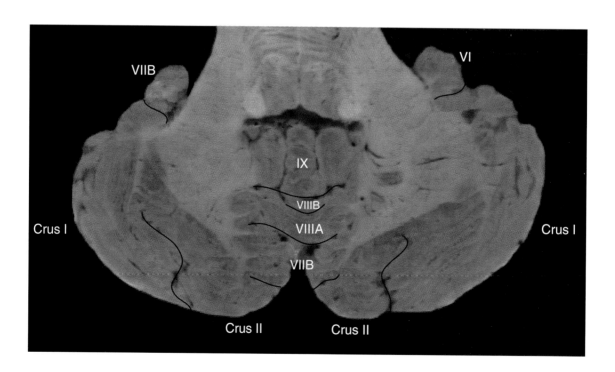

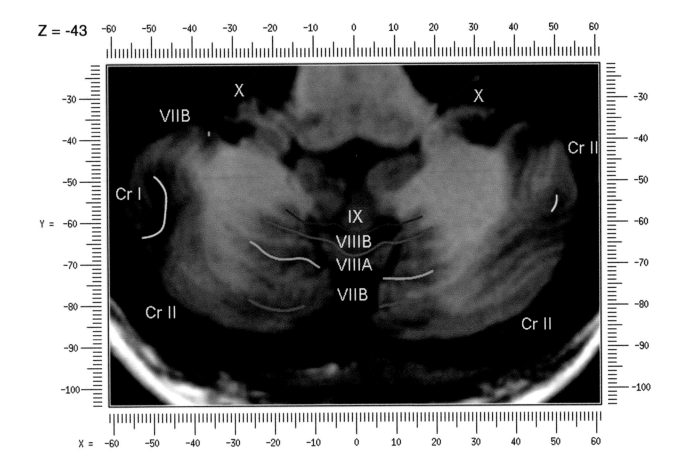

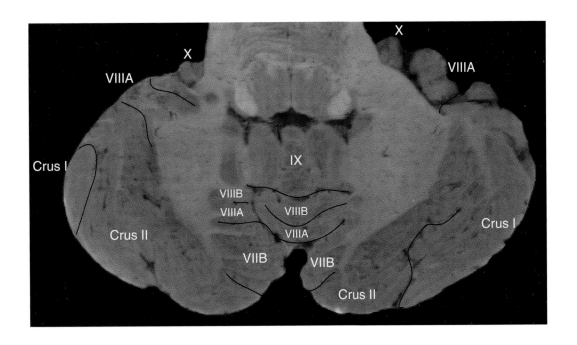

Z = -45

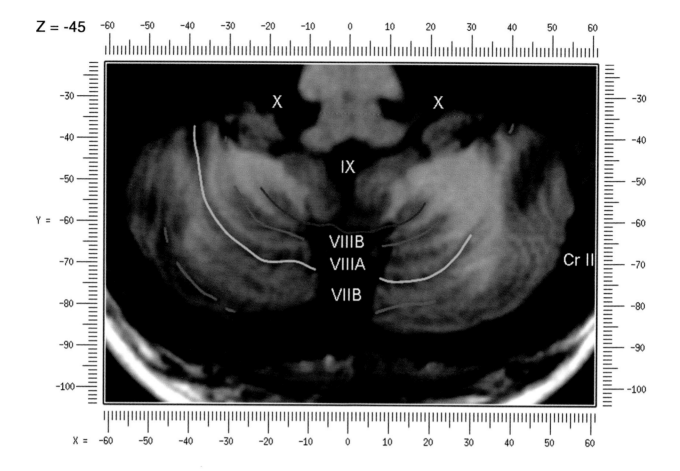

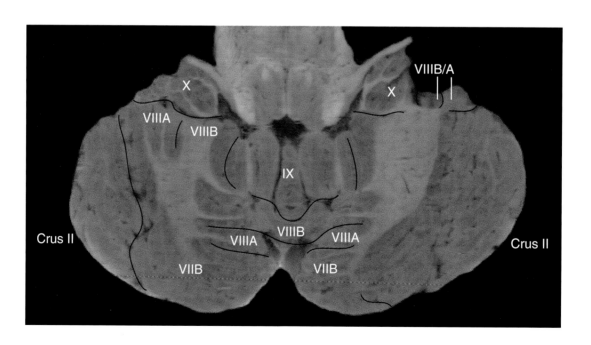

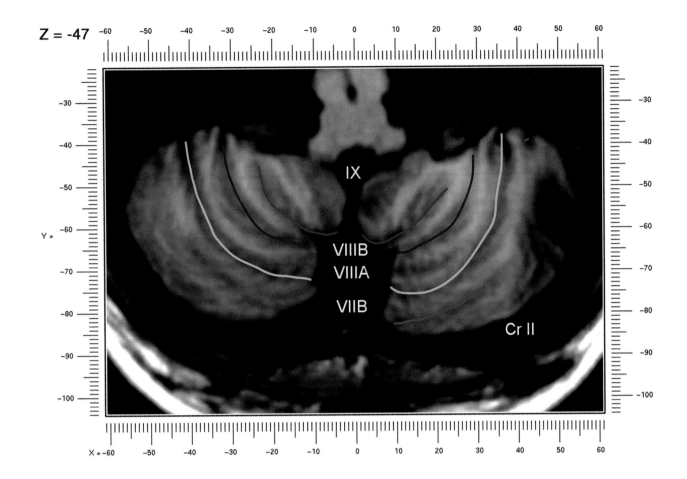

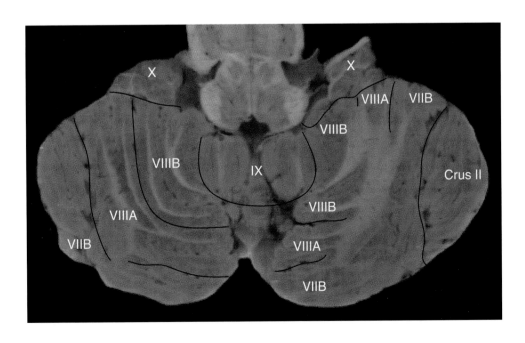

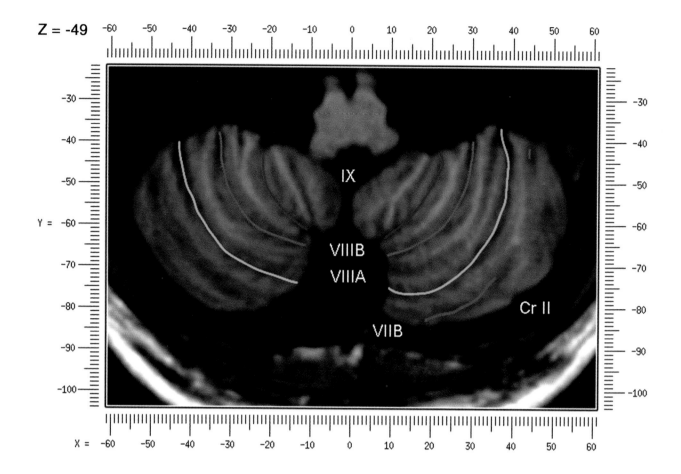

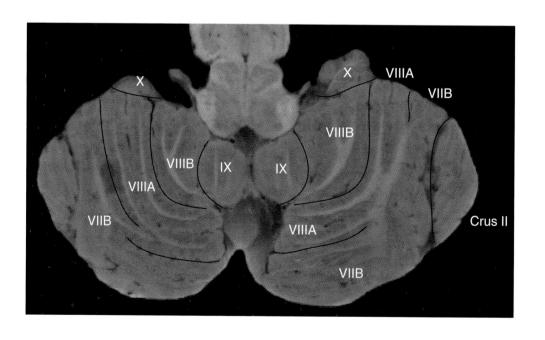

Z = -51

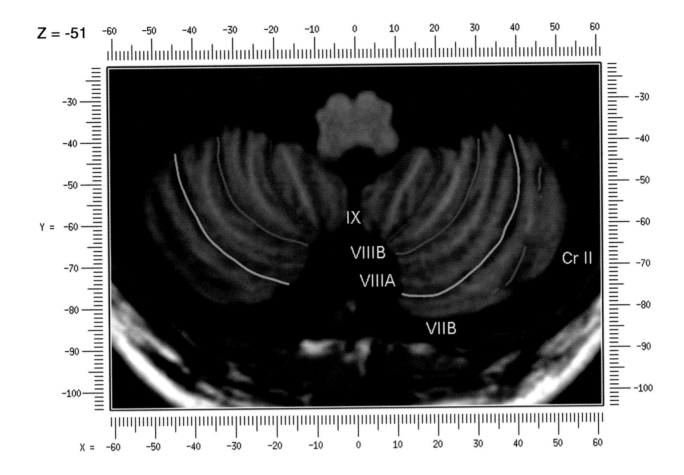

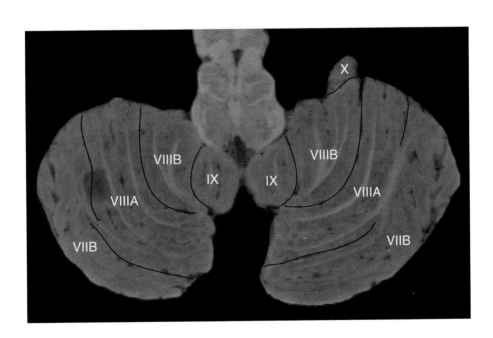

Z = -53

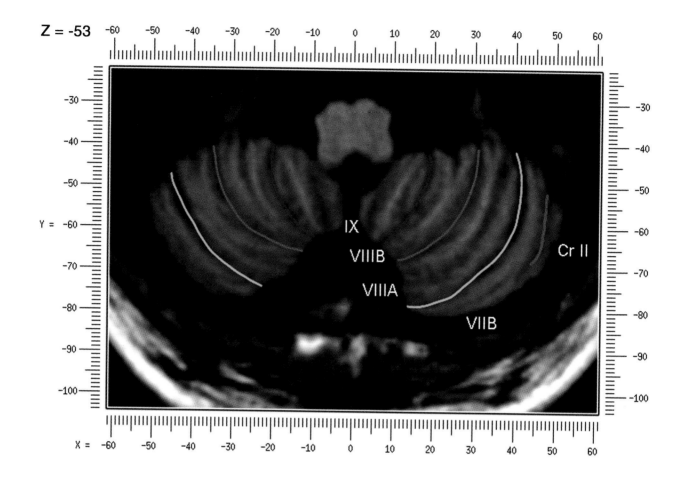

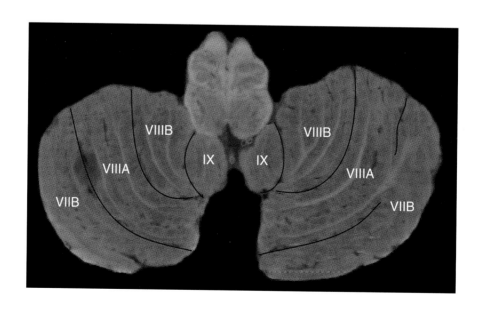

Z = -55

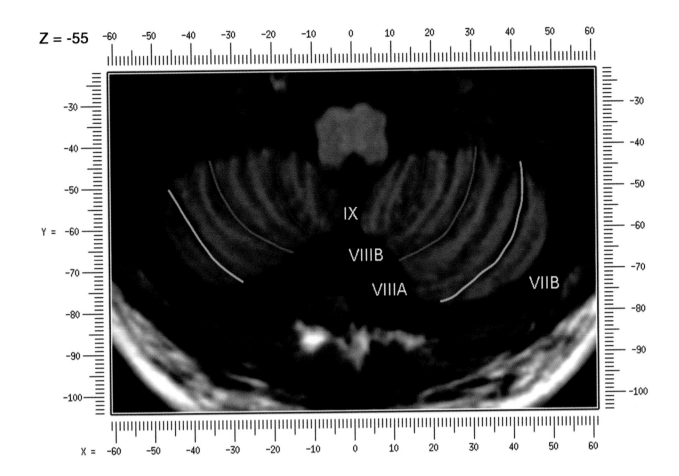

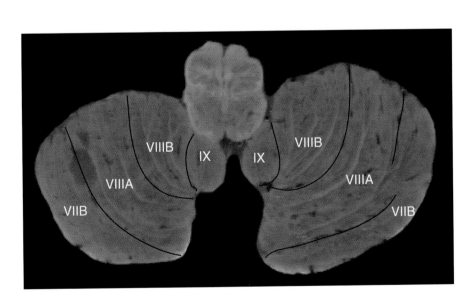

Z = -57

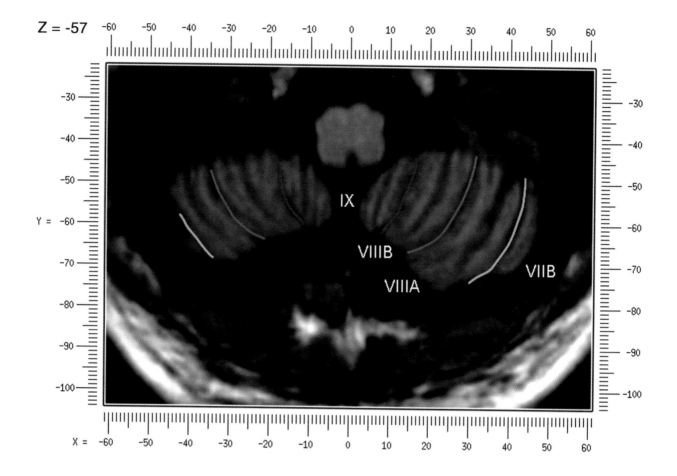

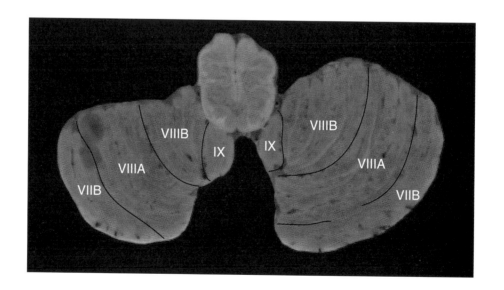

Z = -59

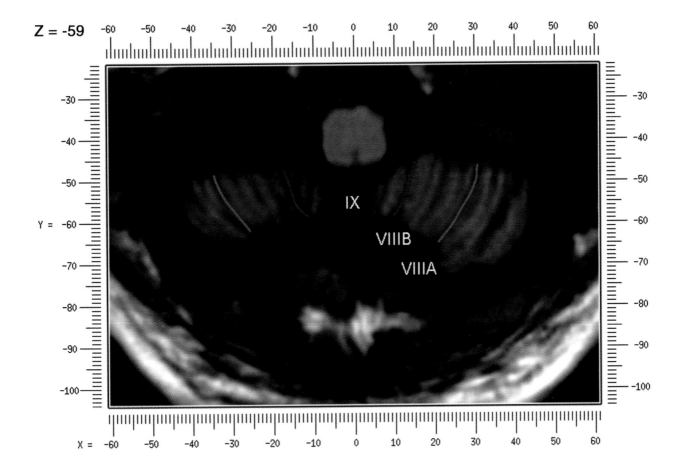

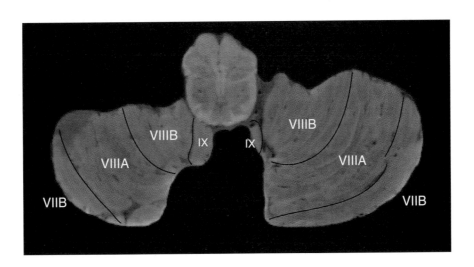

How To Use This CD-ROM

This CD contains a PDF (Portable Document Format) version of the front matter and Introductory Text and low-resolution JPG versions of the Images portion of the *MRI Atlas of the Human Cerebellum*.

To view the PDF, you must have Adobe Acrobat or Adobe Acrobat Reader installed on your computer. If you do not already have one of these applications, you can install it by double-clicking the Acrobat Reader Installer included on this CD.

Once you have Acrobat Reader installed, simply double-click the MRI.pdf icon to view this document.

The three main folders containing the images represent Sagittal, Coronal, and Horizontal sections of the book. In each of these main folders, you will find the images separated by folders for the Cryosection images, MRI images, and Myelin and Nissl stains.

In each of these subfolders, the images are named by page number (022 through 167), location on page (1 for top, 2 for bottom, no designation for only image on page), and coordinate interval. For example, the image named 092_1_COR-32.jpg would be found in the book on page 92, top image on the page, Coronal section, coordinate interval –32.

The images on the CD are intended for instructional use only. These images are not print quality; if printed there will be a lesser degree of detail than the printed images in the book.

The purchaser is entitled to make archival copies of the CD for personal use only. Any other redistribution or copying of this CD is strictly prohibited.

Adobe Acrobat System Requirements

Windows

- P120 Pentium® processor-based personal computer
- Microsoft® Windows® 95 or later
- 32 MB of available RAM
- 10 MB of available hard disk space

Macintosh

- Apple Power Macintosh or compatible computer
- Mac OS software version 7.1.2 or later
- 32 MB of available RAM
- 8 MB of available hard disk space

Technical Support

For technical support, contact the Harcourt Technical Support Center at the numbers indicated below. Service is available in English only between the hours of 7 AM and 6 PM US Central Time (15:00 to 02:00 GMT), Mondays through Fridays.

Toll free in the US and Canada	877-809-6433
Direct dial	817-820-3710
Toll free fax in the US only	800-354-1774
Direct fax	817-820-5100
E-mail	tscap@hbtechsupport.com